Nanoscience Education, Workforce Training, and K-12 Resources

Nanoscience Education, Workforce Training, and K-12 Resources

Judith Light Feather and Miguel F. Aznar

CRC Press
Taylor & Francis Group
Boca Raton London New York

CRC Press is an imprint of the
Taylor & Francis Group, an **informa** business

We would like to thank and credit the following people for providing photographs for the cover:

Phillip Hoy, of students engaging in hands-on activities at the Pennsylvania State University Nanotech Academy (top middle, top right, and bottom left).

Miguel F. Aznar, of students using NanoEngineer-1 at the COSMOS nanotechnology course at the University of California at Santa Cruz (middle left).

Amy Brunner, of students assembling a vacuum system during the Capstone Semester at the Pennsylvania State University Center for Nanotechnology Education and Utilization, and Nanofabrication Facility (bottom right).

CRC Press
Taylor & Francis Group
6000 Broken Sound Parkway NW, Suite 300
Boca Raton, FL 33487-2742

Printed in the United States of America on acid-free paper
10 9 8 7 6 5 4 3 2 1

International Standard Book Number: 978-1-4200-5394-4 (Paperback)

Library of Congress Cataloging-in-Publication Data

Feather, Judith Light.
Nanoscience education, workforce training, and K-12 resources / Judith Light Feather and Miguel F. Aznar.
p. cm.
"A CRC title."
Includes bibliographical references and index.
ISBN 978-1-4200-5394-4 (pbk. : alk. paper)
1. Nanoscience--Study and teaching (Elementary) 2. Nanoscience--Study and teaching (Secondary) 3. Elementary school teachers--In-service training. I. Aznar, Miguel F. (Miguel Flach), 1964- II. Title.

LB1585.F37 2011
620'.5071--dc22 2010023554

Visit the Taylor & Francis Web site at
http://www.taylorandfrancis.com

and the CRC Press Web site at
http://www.crcpress.com

Contents

Section IV Framework Applied

Foreword

James S. Murday
University of Southern California
Office of Research Advancement
Washington, DC

The need to improve U.S. student performance in science, technology, engineering, and mathematics (STEM) education is well documented, with five recent reports as examples:

- The National Academy of Sciences (NAS), "Is America Falling Off the Flat Earth?"[1]
- The National Science Board (NSB), "National Action Plan for Addressing the Critical Needs of the U.S. Science, Technology, Engineering and Mathematics Education System"[2]
- The NSB, "Moving Forward to Improve Engineering Education"[3]
- The Carnegie Corporation of New York, and the Institute for Advanced Study, "The Opportunity Equation: Transforming Mathematics and Science Education for Citizenship and the Global Economy"[4]
- The National Research Council (NRC), "Engineering in K-12 Education: Understanding the Status and Improving the Prospects"[5]

With the disruptive discoveries already realized through nanoscale science and engineering research, it is essential to examine what impact the nanoscale might contribute to approaches to revamp and revitalize STEM education:

- **New knowledge:** Nanostructures can have new physical, chemical, and biological properties. This new knowledge should be incorporated into the educational corpus. There are currently well over seventy thousand nanoscale science and engineering research papers published every year that are rapidly pushing the frontiers of our knowledge base.
- **Transdiscipline:** Nanoscale science and engineering is largely transdisciplinary. It challenges the traditional science and engineering

education taxonomies, which tend toward narrowly scoped courses on biology, chemistry, and physics. To highlight this problem, even before the nanoscale advances in science and engineering, it was already hard to teach modern biology without an appreciation of organic chemistry.

- **Societal impact:** The nanoscale offers sufficient novelty to attract students to STEM, especially since there are numerous examples of its growing impact on their lives. Nano-enabled technologies will contribute toward the solutions of many of our critical societal problems in information management, renewable energy, energy conservation, potable water sources, environmental protection and remediation, and medicine/health.
- **Workforce training:** As nanostructures become materials building blocks and directed self-assembly becomes a viable manufacturing process, there will be a need for an informed, skilled workforce. Various estimates of the commercial market put the value of nano-enabled technologies over $1T/year within the coming decade. The manufacture of those technologies will need a skilled workforce.
- **Risk management:** Workers and members of the general public may be in contact with nanomaterials in various forms during manufacture or in products and should be sufficiently knowledgeable to understand the benefits and risks.

This book addresses the problems the United States faces in revamping its educational system, with the nanoscale providing specific examples to illustrate the challenges and potential solutions. Section I sets the context by examining both U.S. education and nanoscale science/engineering from a societal perspective. Section II then introduces approaches to incorporating the new nanoscale information into our various STEM educational resources. Section III provides guidance toward the many new resources that have already been created, mostly by universities/institutions funded by the National Science Foundation, but also globally. However, those resources are limited, not fully developed, and underutilized. Section IV looks at various options to better exploit the nanoscale toward a new education paradigm. Revamping U.S. STEM education is a daunting task; this book provides new perspectives in how to accomplish that task.

Endnotes

1 "Is America Falling Off the Flat Earth?" The National Academies, Norman Augustine, Chair, Rising Above the Gathering Storm Committee. ISBN 0-309-11224-9, 2007.

2 A National Action Plan for Addressing the Critical Needs of the U.S. Science, Technology, Engineering, and Mathematics Education System. National Science Board, NSB-07-114, Oct 30 2007. http://www.nsf.gov/nsb/documents/2007/stem_action.pdf
3 Moving Forward to Improve Engineering Education, National Science Board, Nov 19 2007, NSB-07-122.
4 The Opportunity Equation: Transforming Mathematics and Science Education for Citizenship and the Global Economy, Carnegie Corporation of New York and Institute for Advanced Study, 2009, www.OpportunityEquation.org
5 Engineering in K-12 Education: Understanding the Status and Improving the Prospects, Katehi, L., Pearson, G., and Feder, M., eds., National Research Council, ISBN 978-0-309-13778-2, 2009.

Preface

Over the past 15 years I have been invited to work with many educational organizations concerned with the lack of improvement in student science achievement tests in grades K–12. The study of science has often been compartmentalized into topics without sequence or depth, followed by testing of a few facts which may not be retained. This disconnected approach to science has left many of our students bored and confused, while teachers have been required to teach material for national and state testing.

Now a new scale/size of science has been introduced to the world, creating excitement among scientists and educators. The nanoscale of science holds so much promise that governments worldwide have developed funding initiatives to explore the research potential leading to new enabling technologies defined as nano, bio, cogno, and info. This small size of scientific research allows mankind to move and manipulate atoms while studying nature to understand self-assembly "from the bottom up" in the hopes of imitating it through technology.

A predominance of cognitive researchers in the United States, who have been funded to study how children learn, have expressed opinions that all teachers need special training to introduce nanoscience in K–12 classrooms. Meanwhile, other countries are including teachers in the development of curriculum materials and successfully introducing nanoscience to K–12 students without special training. Ultimately these decisions in the United States will be made "from the top down" by policymakers who continue to fund the research to study how children learn. Unfortunately, many stakeholders feel it will be decades before nanoscience education reaches classrooms under our present system, and our students will lag behind for decades, handicapped in their ability to compete globally.

This book presents an overview of the current obstacles that must be overcome within the complex U.S. education system before any reform is possible. Examples of inspired students who step forward and make a difference in their schools are provided as incentives to students, parents, and teachers to take an interest and participate in public schools at the local level. Creative ideas for team teaching combining science, art, language, and writing are presented for your consideration. Many public schools are attempting to update their classrooms with technology to expand the choices of instructional materials to include e-learning interactive visual elements. We encourage teachers and students to explore these resources together and change the system "from the bottom up."

Science is the study of nature, and nanoscience is the study of nature at the nano scale—a newly explorable size that opens the window to atoms, molecules, and cellular structures that can be viewed with online microscopes.

It is our hope that you find the book useful and embrace the ideas of moving your classrooms into the twenty-first century. The students are not waiting for permission to learn.

Section I, "Foundations," addresses the national educational matrix starting with an introduction to the scientific and social implications regarding the delay in adopting nanoscience education in public schools. The history of the Department of Education and its mandated structure expands our understanding of how the system operates by law. The structures are explained from the national level to the local school boards, defining the parameters of public education. Successful programs initiated by students are shared as examples of positive change. The section ends with a personal look at three Nobel Prize laureates in physics and chemistry who challenged incompetency and complacency in the education system.

Section II, "Teaching Nanotechnology," turns to the critical process of teaching K–12 students the skills to understand and evaluate emerging technologies they will encounter in the future. Written by Miguel F. Aznar, Educational Director of Foresight Institute, who teaches technology courses for the COSMOS summer school program for high school students at the University of California, Santa Cruz, this section reaches out to teachers by defining a new program on how to teach students to understand and evaluate various emerging technologies.

Section III, "Nanoscience Resources and Programs," reviews the resources of funded outreach programs from universities with Nanoscience Centers. It includes four chapters of categorized resources, teacher development programs, summer camps, and nanoscience educational materials for K–12 in the United States and globally. The section has an overview of current state programs for nanotechnician workforce training as an important new career path for high school graduates.

Section IV, "Framework Applied," is an overview of structure from the national government programs and skill level recommendations for nanoeducation from the National Nanotechnology Initiatives. Also included are the findings and proposed recommendations from the *Workshop for Partnerships in Nanoeducation,* a national stakeholders meeting held in 2009. The multiple levels of education standards for curriculum development are provided with links for educators. New education programs announced by President Obama for 2011 are included, closing with a Senate bill for nanotechnology workforce education submitted on March 15, 2010.

The section also provides a chapter on development tools and supportive Web sites for teachers to explore, and ends with a chapter supporting teachers' efforts to rethink and shift the education paradigm in public schools. The final inspiring discussion highlights a K–12 educational program developed by teachers in Taiwan, outlining their methodology for teaching nanoscience to young students. The chapter closes with questions for thought, as teachers are encouraged to use the tools and resources to introduce nanoscience in their classrooms by exercising the "bottom-up" approach to education decisions.

Authors

Judith Light Feather

President, The NanoTechnology Group Inc.

My early interests in education for K–12 were initiated in 1995 when I worked for a Space Information Group and was able to evaluate some of the educational projects at Johnson Space Center (NASA JSC). During this time period I was asked to join a think tank to discuss avenues of future space colonization and prepare a direction that would investigate aspects of spirituality, integrity, and ethics for colonization of other planets. This direction allowed us to look at curriculum development and the current status in education for a future generation that would actually live and work in space.

Soon after the think tank dissolved, I was introduced to Paul Messier from the National Learning Foundation in Washington, DC, which was exploring skill levels necessary for the future workforce based on changes in technology for education. The Internet had just been introduced, and the military was already using simulations for training with adaptive engines. This led to consultant work as their project coordinator to take the Agile Skills idea and develop a matrix for twenty-first century education utilizing the Internet for e-learning. Funding was not forthcoming because we were ahead of the curve, with only 14% of schools nationally connected to the Internet.

In 1998, I was introduced to Michael McDonald, Executive Director for the NanoComputer Dream Team Inc., who also worked with NASA JSC. He needed an executive assistant to develop a Gold Team for the future of nanoscience education in grades K–12. Rice University had recently been funded for the Center for Biological and Environmental Nanotechnology, which became our first university member to explore this new size of science. By 2002, the Gold Team grew in membership, and we formed The NanoTechnology Group Inc., as a Global Education Consortium to facilitate nanoscience education in K–12.

This growth drew inquiries from Milind Pimprikar in Canada, who was interested in exploring nanotechnology for space applications. Using the same matrix, we worked together to found the CANEUS Organization as a Global Consortium that included space agencies from Canada, Europe, the United States, and Japan. They also became a supporting member of our group with a goal of expanding global education and skill levels necessary for space applications and the future of space colonization.

Our work reached global proportions immediately, and by 2003, I was invited as the keynote speaker in Thailand for the 1st Nano Education Conference for Human Resources. In 2005, I was invited to participate in the Expert Working Group for Nanotechnology in Developing Nations with the United Nations Industrial Development Organization and the International Centre for Science and High Technology (UNIDO ICS), in Trieste, Italy. From 2002 to 2008, I accepted invitations to review emerging technologies in Switzerland, producing overviews of micro/nano, environmental, aerospace, defense, homeland security, and micromechanical technologies. Visiting the research labs of universities and corporations across Switzerland expanded my comprehension of the various aspects from education to commercialization of the integrated fields of science.

In 2007, I was nominated by Akhlesh Lakhtakia, Ph.D., Penn State University for the Harold W. McGraw International Award for Global Awareness in Education. I am presently serving on the Board of the NanoEthics Group, and multiple boards at Lifeboat Foundation.

The NanoTechnology Group Inc. maintains two Internet sites: www.TNTG.org provides resources for every aspect of nanoscience education from curriculum to development tools for teachers. The News Division provides informal educational information and news for the public at www.NanoNEWS.TV

Miguel F. Aznar

Director of Education,
Foresight Institute

I have long been fascinated by how we understand and evaluate technology. The patterns underlying our tools became clear over years, starting even before I studied electrical engineering and computer science at UC Berkeley and continuing through the 1998 founding of KnowledgeContext, a 501(c)(3) nonprofit corporation that helps young people think critically about technology. KnowledgeContext has attracted a diverse team of teachers, technologists, and businesspeople to develop curriculum on understanding and evaluating technology. Through its Web site, KnowledgeContext has provided that curriculum to well over a thousand teachers and now also offers it in wiki form, enabling open collaboration on new versions of the curriculum.

The curriculum is based on a simple strategy for understanding and evaluating technology. In 2004, I brought that strategy to COSMOS, a summer program for mathematics and science at the University of California at Santa Cruz. There may be no better argument for the importance of understanding and evaluating our tools than the power of nanotechnology. While we may be content with simply knowing how to operate many of our technologies, the extraordinary costs and benefits—both existing and promised—of nanotechnology make clear that the future of civilization depends on collaborative, context-aware, critical thinking. With this realization, in 2005 I focused the strategy on nanotechnology, using it to structure an intensive course for precocious high school students in the COSMOS program, which I have been teaching and refining since.

Coincident with creating my nanotechnology course in 2005, I joined Foresight Institute as director of education. The Foresight Institute is the leading think tank and public interest institute on nanotechnology. Founded in 1986, Foresight was the first organization to educate society about the benefits and risks of nanotechnology. For a teacher of nanotechnological literacy, Foresight was an obvious fit.

The concept of technological literacy, as opposed to technological competency, is new to many teachers. It may be that the rapid technological change we are experiencing encourages a myopic view, narrowed to how we operate computers. But that means missing the big picture of understanding and evaluating any technology, from stone tools and printing presses to biotechnology and nanotechnology. Repeat encounters with teachers who were confused by the distinction between literacy and competency encouraged me to write a book. Although on a subject I had been living and breathing, it took years of research to uncover the patterns that transcend specific technologies and allow nonexperts to comfortably operate at a level of abstraction. In 2005, I published *Technology Challenged,* a book that draws from around the world and from the beginning of civilization to reveal patterns in our tools and to offer a simple strategy for anyone to understand and evaluate any technology. Although intended to entertain with stories from Hawaiian bobtail squid to Australian aborigines, the book has also been adopted as a text by several colleges.

I serve as executive director of KnowledgeContext and on the advisory boards of both the Nanoethics Group and the Acceleration Studies Foundation. I have presented at educational conferences, including Computer Using Educators (CUE), California Educational Research Association (CERA), and California League of Middle Schools (CLMS). I have keynoted educational conferences, including Consortium for Research on Educational Accountability and Teacher Evaluation (CREATE), Preparing Tomorrow's Teachers for Technology (PT3), and California Middle Grades Partnership Network (CMGPN). In November of 2006, Google invited me to present a Tech Talk on technological literacy (http://video.google.com/videoplay?docid=-8915225721798779498).

Prior to entering education, I was in management consulting (Ernst & Young, AT&T) and software engineering (Amdahl, Open Systems Development). I was Phi Beta Kappa at UC Berkeley and presided over the Tau Beta Pi engineering honors society while studying there.

Contributors

Amy Brunner
Pennsylvania State University
University Park, Pennsylvania

Robert D. Cormia
Foothill College
Los Altos Hills, California

Dominick E. Fazarro, Ph.D., CSTM
Department of Human Resource
Development and Technology
The University of Texas
Tyler, Texas

Anita Goel, M.D., Ph.D.
Nanobiosym, Inc
Nanobiosym Diagnostics, Inc
Cambridge, Massachusetts

Phillip Hoy
Pennsylvania State University
University Park, Pennsylvania

Akhlesh Lakhtakia, Ph.D., D.Sc.
Pennsylvania State University
University Park, Pennsylvania

James S. Murday
University of Southern California
Office of Research Advancement
Washington, D.C.

Deb Newberry, MSc.
Dakota County Technical College
(DCTC)
Rosemount, Minnesota

Anne Osbourn
John Innes Centre
Norwich, UK

Walt Trybula, Ph.D.
Texas State University–San Marcos
San Marcos, Texas

Introduction

Transformative technologies surgically alter familiar landscapes. Rhythms of life are disrupted and replaced. As old orders yield to new ones, those who profit most are the ones who cheerfully cope with, and even celebrate, transformations.

Within the last decade and a half, nanotechnology has been recognized as transformative. Simultaneous huge advances in biotechnology and information technology, coupled with rapid strides in cognition science, herald a new era. To whom it will bring prosperity and to whom it will bring poverty, within the next decade or two, cannot be foretold with certitude. But clearly, those who will not cope well with the emerging transformation will be relegated to the slag of history, smoldering for a while before going cold.

That would be a great pity, because global problems—such as climate change, massive extinction of species, and widespread microwarfare called terrorism—are so intense that every available mind must be harnessed to overcome their challenges to sustainable ecosystems in which humans continue to play major roles. We have some six billion minds, but the vicissitudes of life have kept most of them functioning at small fractions of their potentials. Worse may happen if backward slides were to occur amidst current concentrations of highly functioning minds.

Any successful strategy to cope with the emerging transformation must involve right education of our children and grandchildren in grade schools. Curriculums today are generally of the just-in-case type: students are taught certain subjects at certain levels for a certain period, with the understanding that some of the students may someday have to use some part of the acquired knowledge. Some curriculums are horizontally integrated, others vertically integrated. A large percentage of students in some countries remain unfamiliar with major themes in mathematics and science. An overwhelming emphasis on those subjects stunts artistic, athletic, and civic development in other countries.

But the grade-school scenario is not bleak at all, for there is much merit in current curricular practices. Acquisition of a broad background is a hedge against an uncertain future for any specific student. Moreover, the ranks of educators embracing collaborative learning and active learning continue to swell. But successfully coping with the transformation emerging due to nanotechnology, information technology, biotechnology, and cognition science requires supplementation by a different instructional approach.

This supplemental approach is just-in-time education (JITE).[1] Well-suited to address complex issues and to solve multidisciplinary problems, JITE is envisaged in terms of experiences, each of which is a project that spans at least two but preferably more scientific and mathematical disciplines. A project may be undertaken by a single student or a team of students, as appropriate, and every student must undertake single-member as well as team projects. The hallmark of JITE is that students will learn to identify the disciplines intersecting a complex problem; to acquire the necessary pieces of information and understanding from each intersecting discipline; to synthesize the various parts into a whole that denotes an acceptable, if not desirable, level of accomplishment; to assess requirements for further developments; and to establish the values of their accomplishments in the cultures of their surroundings, nation, and the world.

Organization and communication skills will be acquired; when in a team project, individuals will be apportioned specific tasks, whose completion will have to be reported to the team before certain deadlines. Tasks and reporting deadlines will also be delineated for single-member projects. Crucially, only a part of the necessary information will be imparted to the students in regular coursework, the remainder to be gathered from untaught portions of schoolbooks, extracurricular books, the Web, site visits, and interviews with practitioners. Different teams undertaking the same project will be encouraged to arrive at different conclusions and deliverables.

Introspection and reflection constitute another crucial aspect of JITE. The value of project tasks to the student will be assessed by him/her before and after undertaking each task. Making use of a daily diary, every student will submit a statement of personal growth: what he/she had expected during the initial stages of the project, and what was actually learnt by the end of the project. Moreover, the statement will contain reflections on the relevance of the project to the town, province, nation, and the world; enhancement of cultural and ecological diversity and sustainability; and suggestions for follow-up projects and other activities.

Thus, JITE is envisaged to impact the teaching and learning not only of science and mathematics, but also of humanities and social sciences. JITE experiences will have to be guided by teams of teachers drawn from a diverse array of disciplines encompassing language arts, sociology and history, civics and political science, physics, chemistry, biology, and mathematics. Science and mathematics teachers will have to learn humanities and social sciences, but even more importantly, humanities and social science teachers will have to learn mathematics and sciences. Finally, only teachers who are themselves lifelong learners shall be able to effectively turn their students into lifelong learners.

In this book, Judith Light Feather and Miguel F. Aznar present a variety of resources for schoolteachers interested in JITE for nanotechnology. Ways to integrate science, mathematics, humanities, and the social sciences are outlined along with descriptions of attempts at various institutions worldwide. The book is both informational and instructional, besides being inspirational and easy to read.

Akhlesh Lakhtakia
The Charles Godfrey Binder (Endowed) Professor of
Engineering Science and Mechanics
Pennsylvania State University

References

1. A. Lakhtakia, Priming pre-university education for nanotechnology, *Current Science* 90(1), 37–40, 2006.

Section I

Foundations

I am a firm believer in the people. If given the truth, they can be depended upon to meet any national crises. The great point is to bring them the real facts.

Abraham Lincoln

1

Introduction to Nanoscience, Technology, and Social Implications

Look deep into nature, and then you will understand everything better.

Albert Einstein, Physicist

Albert Einstein uttered this prophetic statement decades before we were gifted with the ability to work with nature at the nanoscale of science. Advanced microscopy now allows researchers to manipulate and move atoms while searching for answers to the mystery of self-assembly at the atomic scale of matter. Opening this window into the mysteries of our natural world with nano-enabled technologies could solve clean water and accompanying sanitation issues, enable clean energy solutions, and solve health problems to the benefit of humanity.

Inclusion of Nanoscience Education in Schools Is Important for Students

Because science is basically the study of "how the world works" from the subatomic scales to the immense scale of cosmology, addressing our education goals with this in mind benefits all students. The study of science should flow along a line from primary reality to functional truths and then proceed to practical applications that can be used in real-world situations. Interweaving the main scientific disciplines of physics, chemistry, and biology as connected processes interacting as networks throughout nature builds a strong foundation for the study of advanced science. It is often said that nature is our teacher at the nanoscale, helping us gain a deeper understanding of the patterns and relationships that allows even young students to "connect the dots."

Implementing new strategies and solutions in our K–12 classrooms is imperative to address workforce development issues that already reach beyond the laboratory. In order to accomplish this objective, the stakeholders must seriously consider a paradigm shift in how we educate our children. A systems thinking approach to cognition—as a process of knowing—leads to a new method of sharing information. In the theory of living systems described by Fritjof Capra in *The Web of Life*,[1] he states, "Mind is not a thing, the brain is a structure through which the process of cognition operates. Mind and matter are now merely different aspects or dimensions of the phenomena of life." This process of life is based on interactions with the environment. To study science with this approach, we must speak the language of nature, which is based on patterns and relationships. This shift in perspective moves education from rote memorization of facts that match accountability testing requirements to a stimulating, innovative platform capable of engaging and challenging students' minds. It is time to open this window of nature to our children and let them glimpse a future based on nano-enabled technology. Society needs the ability to identify, understand, and evaluate emerging technology from the divergent fields of nano, info, cogno, and bio as it streams into the marketplace.

Detailed Roadmap for the Twenty-First Century[2]

To gain a better understanding of the relevance of this century's technological advances to education, take a look at the project created by Peter Pesti at Georgia Tech College of Computing. Pesti is tracking predictions from visionary/futurist experts on advances in all areas of technology to follow the success or failure of the compilation. He provides a year-by-year bullet point list of notable advances expected to debut in the twenty-first century from 2006 onward. Karsten Staack made the video *21st Century: What will it look like?*[3] using his selections of predictions from the project. Looking ahead for an entire century provides students with a snapshot of technology that resembles science fiction, thus forcing them to consider decisions humanity has never faced. Most of the predicted technology is already in the pipeline of research labs around the world involving human enhancement, implantable devices, DNA and cell modifications, nanobot-targeted drug delivery, environmental solutions requiring modifications of nature, robotic engineered cyborgs (part human/part robots), and advanced artificial intelligence predicted to be smarter than humans before the end of this century.

Understanding the Size in Nanoscience Is a Prerequisite for Teachers

Nano in Greek means dwarf, and is a prefix, so the word nanoscale indicates scale of size, nanosecond indicates time, etc. The remarkable properties exhibited in products made with nanomaterials are due to their tiny subatomic scale. Particles at the nanoscale exhibit quantum behavior that is significantly different than behavior defined by standard physics at the macro scale of matter.

Official Definition of Nanoscience and Nanotechnology

In February of 2000, the National Science and Technology Council Committee on Technology, Subcommittee on Nanoscale Science, Engineering and Technology (NSET) derived the following definition for nanoscience and the resulting nanotechnology (and nanomaterials) development.

> Research and technology development at the atomic, molecular or macromolecular levels, in the length scale of approximately 1–100 nanometer range, to provide a fundamental understanding of phenomena and materials at the nanoscale and to create and use structures, devices and systems that have novel properties and functions because of their small and/or intermediate size. The novel and differentiating properties and functions are developed at a critical length scale of matter typically under 100 nm. Nanotechnology research and development includes manipulation under control of the nanoscale structures and their integration into larger material components, systems and architectures. Within these larger scale assemblies, the control and construction of their structures and components remains at the nanometer scale. In some particular cases, the critical length scale for novel properties and phenomena may be under 1 nm (e.g., manipulation of atoms at ~0.1 nm) or be larger than 100 nm (e.g., nanoparticle reinforced polymers have the unique feature at ~ 200–300 nm as a function of the local bridges or bonds between the nano particles and the polymer).[4]

This definition serves as the official platform for government agencies to standardize the development of solicitations with guidelines for proposals, funding, and subsequent research in the integrated fields. It also defines the size for commercialization of products that fall into three categories: materials, devices, and integrated systems.

Size Matters in Scientific Disciplines

The Scale of Things

Examples of nanomaterials and nanodevices abound today in our computers, window coatings, paints, pharmaceuticals, lotions, tennis racquets, golf balls, electronic circuits, catalysts, polymers, and composites.

Figure 1.1, courtesy of the U.S. Department of Energy (DOE), scales and compares natural materials to synthetic ones. The scale begins with objects that are visible and continues down to those that are invisible and and are expressed at the atomic scale.

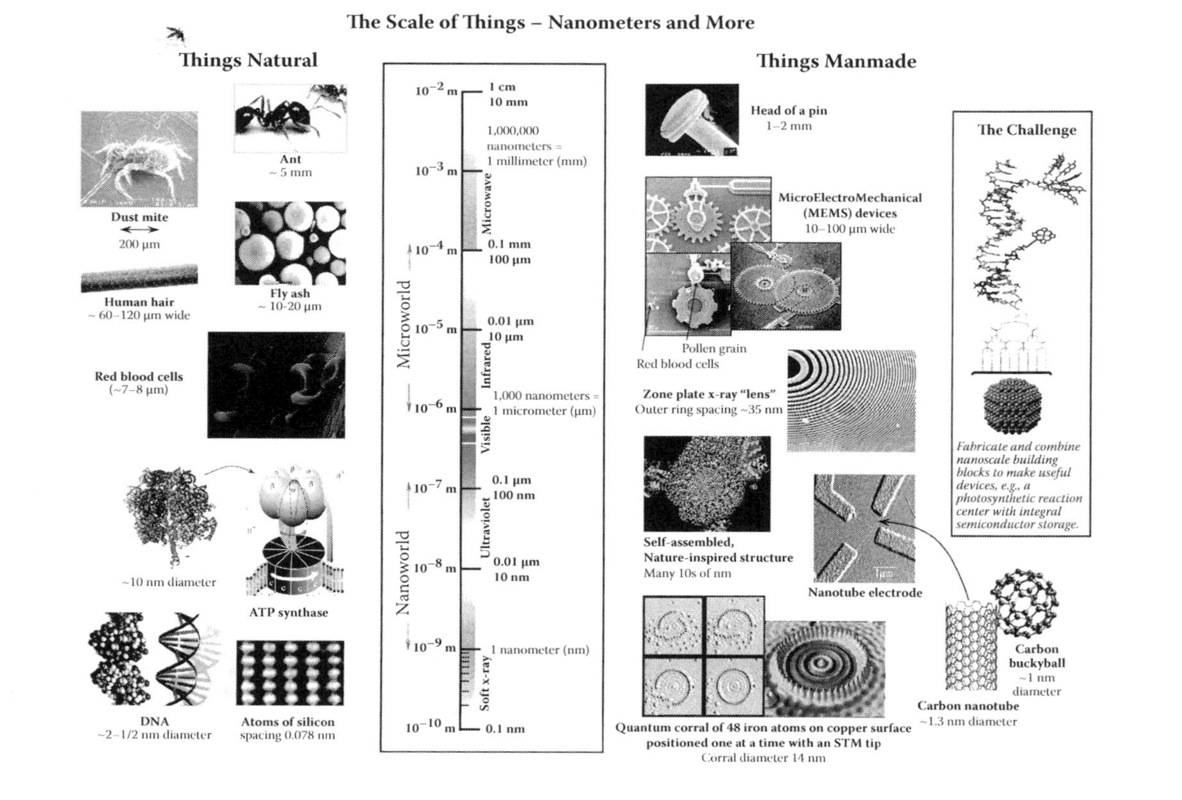

FIGURE 1.1
The above "Scale of Things" chart was designed by the Office of Basic Energy Sciences (BES) for the U.S. Department of Energy using U.S. taxpayers. dollars; therefore, the chart is not copyrighted and may be used without written permission.

The National Nanotechnology Coordination Office has developed a document to enable the public to define nanoscience and nanotechnology titled "Nanotechnology: Big Things from a Tiny World."[5] The document is a wonderful introductory resource for teachers and students. The document guides students to think really small—smaller than anything they can see in a normal microscope at school—down to the size of molecules and atoms. This is the nanoscale, where scientists are learning about these fundamental components of matter, and that knowledge is generating worldwide interest and excitement.

What Is So Special About the Nanoscale?

The short answer is that materials can have different properties at the nanoscale. Some are better at conducting heat and electricity, some are stronger, some have different magnetic properties, and some reflect light better or change colors as their size is changed. A good example is gold, which changes in color to red at the nanoscale.

Nanoscale materials also have a larger surface area, which provides more surface for interactions with the materials around them. The example in the document was a wad of chewed gum stretched to its potential length. The stretching caused the gum to have a larger surface area than the original wad of gum. The stretched gum behaves differently and is likely to dry out faster and become brittle sooner that the wad of gum, which has less surface for the air to affect.

How Small Is a Nanometer?

By definition a nanometer is a billionth of a meter, which is hard to imagine. Some easier ways to think about size is to compare with familiar examples.

- A normal sheet of paper is about 100,000 nanometers thick.
- Blond hair is 15,000 to 50,000 nanometers thick.
- Black hair is 50,000 to 180,000 nanometers thick.
- There are 25,400,000 nanometers to an inch.
- A nanometer is a millionth of a millimeter.

The nanoscale is not new. It is our ability to visually observe atoms and molecules at this size that is new. Nanoscale materials are found all around us in nature. Scientists someday hope to imitate nature's ability to self-assemble and replicate the processes at the nanoscale. Researchers have already copied the nanostructure of lotus leaves that repel water. This process has already been used to make stainproof clothing and other fabrics and materials. Others are trying to imitate the strength and flexibility of spider silk which is naturally reinforced by nanoscale crystals. Humans and animals use natural nanoscale materials, such as proteins and other molecules, to control the body's many systems and processes. In fact, many vital functions of living organisms take place at the nanoscale. A typical protein such as hemoglobin, which carries oxygen through the blood stream, is 5 nanometers, or 5 billionths of a meter, in diameter. Researchers are studying this field of nanobiology to understand these processes and develop new methods of treating disease.

Nanoscale materials are all around us in smoke from fire, volcanic ash, and sea spray, as well as products resulting from burning or combustion processes. Some have been used for centuries. One material—nanoscale gold—has been a component in stained glass and ceramics as far back as the tenth century. But it took 10 more centuries before high-powered microscopes and precision equipment were developed to allow nanoscale materials to be imaged and placed with precision. Therefore, you cannot just throw a bunch of materials together to develop nanotechnology. It requires expanded knowledge of characterization processes and the ability to manipulate and control those materials in a useful way.

Much like following a recipe for baking a cake with multiple ingredients, you cannot just throw it all in a bowl at the same time and turn on the mixer. Separate steps must be taken to prepare sets of ingredients, such as creaming the butter and sugar together before adding the eggs. Dry ingredients such as flour are always added a little at a time after all the other ingredients have been combined. A recipe is a process that is necessary to bake a cake. Learning the processes in materials science and recording your notes during the process creates a final recipe for replication.

However, there is a "bottom-up" approach to making things at the nanoscale, and some researchers are exploring this type of manufacturing for the future. The idea is that, if you put certain molecules together, they will self-assemble into ordered structures. This approach could reduce the waste from current top-down processes that start with large pieces of materials and end with the disposal of excess material.

What about the Behavior of Materials at the Nanoscale?

It is true that materials at this small size behave differently than normal bulk materials. For example, at the bulk scale gold is an excellent conductor of heat and electricity, but nothing much happens when you shine light onto a piece of gold. With properly structured gold nanoparticles, however, something almost magical happens. They start absorbing light and can turn that light into heat, enough heat, in fact, to act like miniature thermal scalpels that can kill unwanted cells in the body, like cancer cells.

Other materials can become remarkably strong when built at the nanoscale. For example, nanoscale tubes of carbon the diameter of a human hair, 50,000 nanometers, are incredibly strong. They are used to make bicycles, baseball bats, and some automotive parts. Combining carbon nanotubes with plastics produces composites far stronger and lighter than steel that are used by the aerospace industry. The composites also conduct heat and electricity, thus protecting airplanes from lightning strikes and cooling their computer circuits. Imagine the fuel savings when these composites are incorporated in the manufacturing of automobiles.

Social Implications

All technologies have impacts on our lives, and along with the benefits, there may be risks. The National Nanotechnology Initiative (NNI) has taken these concerns into consideration from the beginning of their development. Research on environmental, health, and safety impacts and social implications are important areas of funding for the NNI. Funding by the National Institute of Health, National Science Foundation, and the Environmental Protection Agency has increased the knowledge base of researchers to better understand nanoscale materials and to identify any unique safety concerns that may be associated with them. This knowledge guides researchers in development of guidelines for handling and disposal of materials. It also helps them to avoid certain materials in products or to modify the materials to make them safe. According to experts, risk involves two factors: hazard and exposure. Therefore, if there is no exposure even a hazardous material does not pose a risk. Researchers have found that some nanomaterials could provide potential solutions to risks from other technologies and materials. For example, thousands of cases of arsenic poisoning are reported each year worldwide and are linked to well water. Vicki Colvin, Rice University's Kenneth S. Pitzer-Schlumberger Professor of Chemistry and director of Rice's Center for Biological and Environmental Nanotechnology (CBEN), invented an arsenic-removing technology based on the unique properties of particles called "nanorust," tiny bits of iron oxide that are smaller than living cells (nanoscale). In 2006, Colvin and CBEN colleague Mason Tomson, professor in civil and environmental engineering, published with their students the first nanorust studies. Their initial tests indicated nanorust—which naturally binds with arsenic—could be used as a low-cost means of removing arsenic from water.

Social scientists, ethicists, and others are already studying the broader implications of nanotechnology. Identifying positive and possible negative impacts early in the research helps minimize, or avoid, undesirable effects in order to maximize the benefits. Twenty-six departments and agencies of the U.S. government participate in the NNI to coordinate research and development efforts funded by the government.

Most of our universities have been funded to develop education that addresses nanoscale science along with the Nanoscale Centers and National

Laboratories. Many of them have produced student outreach programs and curriculum that teachers and students can access. Section III ("Nanoscience Resources and Programs") provides the most recent listings of all educational materials available for K–12 and workforce training. It is our intent that you find these materials useful and introduce some of them in your classrooms.

The key to our future depends on the successful development, implementation, and inclusion of nanoscience instructional materials in our preuniversity education and workforce training programs for technicians.

References*

1. Capra, F. *The Web of Life*. Doubleday Publishing, New York. ISBN 0-385-47675-2. 1996, Chapter 7.
2. http://www.cc.gatech.edu/~pesti/roadmap/
3. http://www.youtube.com/watch?v=c1KEFgD6Dtg&feature=player_embedded. 2010.
4. Definition, Interagency Subcommittee on *Nanoscale Science, Engineering and Technology* (NSET) of the Federal Office of Science and Technology Policy, February (2000).
5. Nanotechnology: Big Things from a Tiny World http://www.tntg.org/documents/46.html

* All links active as of August 2010.

2

Education Is a Complex System: History, Matrix, Politics, Solutions

> Bureaucracy defends the status quo long past the time when the quo has lost its status.
>
> **Laurence J. Peter (1919–1988)**

The Complexity of Our Education System Is Not Easily Penetrated

The slow but steady decline of education in the United States is well documented,[1] and the results of various studies are released on a regular basis.[2] We are constantly reminded that the quality of education is eroding.[3] However, most studies never explain the complexity of our system that hampers efforts to improve.[4] So how, and where, do we find solutions? When, and how, can new programs be implemented?

This book addresses these vital questions, offering suggestions, options, and resources for further investigation. A journey through the educational matrix to gain insight on how we arrived at this point may help us make thoughtful decisions in preparing our students for their future.

Brief History of Our Education Matrix

Level 1: Policymakers and Legislation from the Top Down[5]

> The original Department of Education was created in 1867 to collect information on schools and teaching that would help the States establish effective school systems. While the agency's name and location within the Executive Branch have changed over the past 130 years, this early emphasis on getting information on what works in education to teachers and education policymakers continues down to the present day.

The passage of the Second Morrill Act in 1890 gave the then-named Office of Education responsibility for administering support for the original system of land-grant colleges and universities. Vocational education became the next major area of Federal aid to schools, with the 1917 Smith–Hughes Act and the 1946 George–Barden Act focusing on agricultural, industrial, and home economics training for high school students.

World War II led to a significant expansion of Federal support for education. The Lanham Act in 1941 and the Impact Aid laws of 1950 eased the burden on communities affected by the presence of military and other Federal installations by making payments to school districts. And in 1944, the "GI Bill" authorized postsecondary education assistance that would ultimately send nearly 8 million World War II veterans to college.

The Cold War stimulated the first example of comprehensive Federal education legislation, when in 1958 Congress passed the National Defense Education Act (NDEA) in response to the Soviet launch of Sputnik. To help ensure that highly trained individuals would be available to help America compete with the Soviet Union in scientific and technical fields, the NDEA included support for loans to college students, the improvement of science, mathematics, and foreign language instruction in elementary and secondary schools, graduate fellowships, foreign language and area studies, and vocational-technical training.

The anti-poverty and civil rights laws of the 1960s and 1970s brought about a dramatic emergence of the Department's equal access mission. The passage of laws such as Title VI of the Civil Rights Act of 1964, Title IX of the Education Amendments of 1972, and Section 504 of the Rehabilitation Act of 1973 which prohibited discrimination based on race, sex, and disability, respectively made civil rights enforcement a fundamental and long-lasting focus of the Department of Education. In 1965, the Elementary and Secondary Education Act launched a comprehensive set of programs, including the Title I program of Federal aid to disadvantaged children to address the problems of poor urban and rural areas. And in that same year, the Higher Education Act authorized assistance for postsecondary education, including financial aid programs for needy college students.

In 1980, Congress established the Department of Education as a Cabinet level agency. Today, ED operates programs that touch on every area and level of education. The Department's elementary and secondary programs annually serve nearly 14,000 school districts and some 56 million students attending roughly 99,000 public schools and 34,000 private schools. Department programs also provide grant, loan, and work-study assistance to more than 14 million postsecondary students.

Level 2: Education from the Top Down Legislates Mandatory Testing for Accountability

We are now at the other end of the education spectrum, where the United States Congress requires national mandatory testing for accountability tied to federal funding programs. The states have a second layer of standards and mandatory testing tied to their funding. These, in turn, are reflected in the textbooks published to match both sets of testing. Due to these "top-down" testing requirements, we are saddled with a national curriculum by default, forcing teachers to use textbooks that teach to the test.

Budget Deficits Are Encouraging Changes in Schools

Due to recent state budget deficits nationwide, many state education departments are investigating cost comparatives concerning technology for classrooms that enables e-learning versus the cost of standard textbooks. An example is California, which announced in 2009 a new Initiative[6] to acquire open source textbooks for the state.

> As California prepares to become the first state in the nation to offer free, open-source digital textbooks for high school students this fall, state officials today gave an A-plus to a North Carolina high school teacher's algebra II textbook, one of the first open-source texts submitted for the program.
>
> Advanced Algebra II[7] by Raleigh, N.C., math teacher Kenny Felder was submitted to California officials by Connexions, an open-education initiative at Rice University in Houston that publishes the open-copyright book.
>
> "Gov. Arnold Schwarzenegger's initiative, together with President Obama's proposal to invest $500 million in open-education over the next decade, are two of the most significant steps forward in open-education to date," said Joel Thierstein, Connexions executive director. "Open education is the biggest advance in education since Horace Mann's push for mandatory free public education in the U.S."
>
> California Secretary of Education Glen Thomas today unveiled his department's review of the first 16 digital texts submitted by publishers in response to Schwarzenegger's May 6 call for free open-source digital textbooks for high school students. Textbook choices are made at the local level in California, and Thomas' reviews are designed to help local officials choose digital books that best meet their needs. The reviews assessed how well each book complied with California's state textbook standards, and Connexions' algebra text scored a 96, meeting 26 of the 27 standards tested.
>
> "One of the beauties of open-education in general, and Connexions in particular, is that anyone who wants to take the time to create content can do it, and anyone who wants to update content and keep it current or improve it can do that too," Thierstein said. "A book is never static in Connexions because everything is published under a Creative Commons Attribution Only copyright license. Any teacher can modify the book to make it culturally relevant for their students."
>
> The reviews of Felder's book and the other submissions for California's K–12 open-source textbook initiative were presented at a symposium in Orange County that was organized by the California Educational Technology Professionals Association. The event attracted hundreds

> of officials who are tasked with choosing curriculum in a year with extremely tight budgets. Thierstein, an invited panelist, answered questions and explained how open-source texts like Felder's book could both improve classroom instruction and save money.
>
> "Everyone is looking to cut costs over the next couple of years, but the real beauty of open-educational resources like Kenny Felder's book is that they provide the foundation for a step-change in the quality of education in the United States," Thierstein said. With more than a million visitors a month and one of the world's largest repositories of open-education resources, Connexions is a leading global provider of open-copyright licensed, free educational materials. Connexions is available free for anyone to contribute to or learn.

The financial crisis may be the catalyst that moves schools into twenty-first century modes of learning. Policymakers are currently tasked to investigate these issues and develop legislative changes to prepare high school graduates with the new skills and tools necessary to be competitive globally.[8]

Following the Funding Trail As It Expands from the Top Down

The Department of Education has kept to its original mandate: only provide research on "how students learn" and "what works in education." The mandate was reinforced on November 5, 2002, as Congress passed the Education Sciences Reform Act of 2002 (ESRA) establishing the Institute of Education Sciences[9] (IES, or the Institute) and its advisory board, the National Board for Education Sciences[10] (NBES, or the Board). The Institute reports to Congress yearly on the state of education in the United States. The IES Web site avers that the Institute provides thorough and objective evaluations of federal programs, sponsors research relevant and useful to educators and others (such as policymakers), and serves as a trusted source of information on "what works in education."

> With a budget of over $200 million and a staff of nearly 200 people, IES has helped raise the bar for all education research and evaluation by conducting peer-reviewed scientific studies, demanding high standards, and supporting and training researchers across the country. We fund top educational researchers nationwide to conduct studies that seek answers on what works for students from preschools to postsecondary, including interventions for special education students. We collect and analyze statistics on the condition of education, conduct long-term longitudinal studies and surveys, support international assessments, and carry out the National Assessment of Educational Progress, also known as the Nation's Report Card.[11] We conduct evaluations of large-scale educational projects and federal education programs—which soon will include examining reforms driven by the American Recovery and Reinvestment Act. We help states work toward data-driven school improvement by providing

> grants for the development and use of longitudinal data systems. Finally, we inform the public and reach out to practitioners with a variety of dissemination strategies and technical assistance programs, including: the What Works Clearinghouse; the ERIC[12] education database; ten Regional Educational Laboratories;[13] national Research and Development Centers;[14] and through conferences, publications and products.
>
> Moving forward, IES' rigorous research agenda will be informed by the voices and interests of practitioners and policy makers, who will be involved in shaping the questions most relevant to their practice. We will seek to build the capacity of states and school districts to conduct research, evaluate their programs and make sense of the data they are collecting. We will strive to develop a greater understanding of schools as learning organizations and study how development, research, and innovation can be better linked to create sustainable school reforms.

In 2002 the Institute (IES) funded $100 million in research and development for the *What Works Clearinghouse*,[15] a Web site that publishes research results for educators, practice guides, and intervention reports. Their most recent report for educators titled: *What works for educators*[16] states the following:

What's in It for You?

> The WWC offers timely, accessible materials to help educators identify and implement research-based practices. Publications and services include:
>
> - Practice guides, which contain explicit suggestions about effective approaches for topics such as organizing instruction and study to improve student learning, reducing behavior problems in the classroom, developing effective out-of-school time programs, assisting students struggling with math and reading, and using data to monitor academic progress.
> - Intervention reports, which are comprehensive reviews of research on educational products, programs, practices, or policies. Topics include reading (beginning reading, adolescent literacy, and learning disabilities), math (elementary, middle, and high school), early childhood education, dropout prevention, and character education.
> - A Help section where you can tour the site, learn answers to frequently asked questions, browse the glossary of terms, and contact a knowledgeable staff member to help you navigate the resources of the WWC.

Visit the WWC and judge for yourself if the original $100 million funding in 2002 was effective, and whether the $200 million a year for 200 employees to write research papers for educators has helped any of the schools or students improve. Also keep in mind that none of this funding is for curriculum development, just research into how students learn.

Following the Funding Trail to the National Science Foundation

Since the Department of Education does not fund curriculum, the next stop for inquiry as to how to include nanoscience instructional materials into the teaching syllabus was The National Science Foundation. In October 2002, the National Science Foundation[17] (NSF) funded centers that focused on research for science, engineering, technology, and mathematics (STEM) education. Again these five new centers were funded to engage in research on how students learn. They were funded to also answer the need for a new generation of professionals who could inspire and challenge students while engaging in research to understand how they learn. The centers are located at: American Association for the Advancement of Science (AAAS) in Washington DC, Washington University in St. Louis, and at the AAAS universities of Wisconsin, Washington, and Georgia according to the NSF press release. Each of those centers was scheduled to receive $10 million over five years for this research. The press release stated that they would continue individual efforts to develop the K–12 component of the program, ranging from the development of new math and science curricula to instructional materials and professional development of teachers. There are a total of ten nationwide K–12 Centers for Learning and Teaching that receive an estimated NSF commitment of $100 million to also increase the numbers, professionalism, and diversity of K–12 math and science teachers, and faculty members who prepare future teachers.

In this same press release, it was stated that, in addition to these centers, two higher education centers were funded by NSF at $20 million to provide coordinated efforts in research, faculty professional development, and education practice at colleges and universities.

These Centers were funded as test sites for innovative approaches in preparing a new generation of science, engineering, and mathematics faculty that could work well together and introduce a strong research component into educational approaches. The total investment for these centers was stated to be $100 million, all basically for research into how students learn. Based on the original mandate for the Department of Education, none of these centers actually develop instructional materials or the final curriculum that is used in the schools. This led us back to the Department of Education to look at the deeper layers of this complexity.

It appears the dichotomy was built into the system from the beginning. This became apparent in the Overview and Mission statement published on the Department of Education Web site.[18]

> Education is primarily a State and local responsibility in the United States. It is States and communities, as well as public and private organizations of all kinds, that establish schools and colleges, develop curricula, and determine requirements for enrollment and graduation. The structure of education finance in America reflects this predominant State and local role. Of an estimated $1.1 trillion being spent nationwide on education at all levels for school year 2009-2010, a substantial majority

will come from State, local, and private sources. This is especially true at the elementary and secondary level, where about 89.5 percent of the funds will come from non-Federal sources.

That means the Federal contribution to elementary and secondary education is a about 10.5 percent, which includes funds not only from the Department of Education (ED) but also from other Federal agencies, such as the Department of Health and Human Services' Head Start program and the Department of Agriculture's School Lunch program.

Mission

Despite the growth of the Federal role in education, the Department never strayed far from what would become its official mission: to promote student achievement and preparation for global competitiveness by fostering educational excellence and ensuring equal access.

The Department carries out its mission in two major ways. First, the Secretary and the Department play a leadership role in the ongoing national dialogue over how to improve the results of our education system for all students. This involves such activities as raising national and community awareness of the education challenges confronting the Nation, disseminating the latest discoveries on what works in teaching and learning, and helping communities work out solutions to difficult educational issues.

Second, the Department pursues its twin goals of access and excellence through the administration of programs that cover every area of education and range from preschool education through postdoctoral research. For more information on the Department's programs see the President's FY 2011 Budget Request for Education.[19]

National Funding for The Condition of Education *Annual Report*

Each June an annual report, also mandated by Congress, is produced by the National Center for Education Statistics (NCES) titled *The Condition of Education*[20] that charts student performance and is published and released free to the public. You will notice in this report, covering 1995–2007, that the United States has no measurable difference in improvement.

Why Are We Not Questioning the Status Quo, When It Is Obviously Not Working?

My question at this point was: why are we still spending upwards of $300 million per year for researchers to figure out how children learn, when the results show no improvement from 1995 through 2007? Why not change the mission of the Department of Education to meet the challenges of the twenty-first century educational paradigm? The mission statement from 1867 does not serve our students' needs in the technological society of 2010; it continues to promote the status quo of failure and nonpreparedness in a global society (Table 2.1).

TABLE 2.1[21]

Trends in Average Science Scores of Fourth- and Eighth-Grade Students, by Country: 1995 to 2007

Grade Four				Grade Eight			
	Average Score		Difference[1]		Average Score		Difference[1]
Country	1995	2007	2007–1995	Country	1995	2007	2007–1995
Singapore	523	587	63*[a]	Lithuania[2]	464	519	55*[a]
Latvia[2]	486	542	56*[a]	Colombia	365	417	52*[a]
Iran, Islamic Rep. of	380	436	55*[a]	Slovenia	514	538	24*[a]
Slovenia	464	518	54*[a]	Hong Kong SAR[3,4]	510	530	20*[b]
Hong Kong SAR[3]	508	554	46*[a]	England	533	542	8
Hungary	508	536	28*[a]	United States[4,5]	513	520	7
England	528	542	14*[a]	Korea, Rep. of	546	553	7*[b]
Australia	521	527	6	Russian Federation	523	530	7
New Zealand	505	504	–1	Hungary	537	539	2
United States[4,5]	542	539	–3	Australia	514	515	1
Japan	553	548	–5*[b]	Cyprus	452	452	[d]
Netherlands[6]	530	523	–7	Japan	554	554	–1
Austria	538	526	–12*[b]	Iran, Islamic Rep. of	463	459	–4
Scotland	514	500	–14*[b]	Scotland[4]	501	496	–5
Czech Republic	532	515	–17*[c]	Romania	471	462	–9
Norway	504	477	–27*[c]	Singapore	580	567	–13[c]
				Czech Republic	555	539	–16*[c]
				Norway	514	487	–28*[c]
				Sweden	553	511	–42*[c]

Note: Bulgaria collected data in 1995 and 2007, but due to a structural change in its education system, comparable science data from 1995 are not available. Countries are ordered by the difference between 1995 and 2007 overall average scores. All countries met international sampling and other guidelines in 2007, except as noted. Data are not shown for some countries, because comparable data from previous cycles are not available. The tests for significance take into account the standard error for the reported difference. Thus, a small difference between the United States and one country may be significant while a large difference between the United States and another country may not be significant. Detail may not sum to totals because of rounding. The standard errors of the estimates are shown in tables E-20 and E-21 available at http://nces.ed.gov/pubsearch/pubsinfo.asp?pubid=2009001.

TABLE 2.1 (Continued)

Trends in Average Science Scores of Fourth- and Eighth-Grade Students, by Country: 1995 to 2007

[*] $p < .05$. Within-country difference between 1995 and 2007 average scores is significant.
[a] Country difference in average scores between 1995 and 2007 is greater than analogous U.S. difference ($p < .05$)
[b] Country difference in average scores between 1995 and 2007 is not measurably different from analogous U.S. difference ($p < .05$)
[c] Country difference in average scores between 1995 and 2007 is less than analogous U.S. difference ($p < .05$)
[d] Rounds to zero.
[1] 1Difference calculated by subtracting 1995 from 2007 estimate using unrounded numbers.
[2] In 2007, National Target Population did not include all of the International Target Population defined by the Trends in International Mathematics and Science Study (TIMSS).
[3] Hong Kong is a Special Administrative Region (SAR) of the People's Republic of China.
[4] In 2007, met guidelines for sample participation rates only after substitute schools were included.
[5] In 2007, National Defined Population covered 90% to 95% of National Target Population.
[6] In 2007, nearly satisfied guidelines for sample participation rates only after substitute schools were included.

Source: International Association for the Evaluation of Educational Achievement (IEA), Trends in International Mathematics and Science Study (TIMSS), 1995 and 2007.

Increased Complexities Hamper Inclusion of Nanoscale Science Curriculum

The complexities increase as we delve into solutions for education. Nanoscience education is particularly difficult to implement in classrooms where students are already failing science in large numbers. Many elementary schools eliminated science during the past decade because it was not included in testing for grades K–8 under the "No Child Left Behind Act." It may take another decade for nanoscience education to be included in the national standards and skill levels "from the top down," so our next step is to look at how we can make changes in our schools "from the bottom up."

Change Happens from the Local School Board Level from the Bottom Up

Parents and communities may not realize the importance of their participation during elections of their local school boards. A recent study by David Webber at University of Missouri–Columbia[22] examined this link between school board elections and local school performance. He found a correlation between increased voter turnout for school board elections and increased state assessment scores. Webber, associate professor of political science at Missouri University College of Arts and Science, questioned whether parental involvement in voting for school board members makes a difference in the test scores. The premise was that voting for local candidates establishes social capital in a school district. Though they could attract citizen involvement based on potential, the study found few candidates vying for school board seats and low voter turnout in most districts.

The study examined election records in 206 Missouri school districts from 1998 to 2001, with voter turnout at only 22%. They also discovered that a 1% increase in voter turnout correlated to an increase in the state assessment scores by more than one point. The study also showed that school districts with lower graduation rates draw a much higher percentage of voters and had more competition for the board seats.

The results concluded that the importance of community involvement in both competing for school board seats and voting are not clearly understood by the public, suggesting educational forums for stimulating more participation. Parents and teachers are on the bottom rung of the ladder of hierarchy in the communities for curriculum in public schools, as shown in Figure 2.1. As we work our way up through the matrix, states make the decisions on standards and pass them down to the districts and school boards.

Level 3: States Collaborate to Develop New Reading and Math Standards

Last year the National Governors Association (NGA) and the Chief State School Officers (CCSSO) were tasked to develop and implement new reading and math standards that build toward college and career readiness. Many states will adopt these standards, proving the governors' initiative was an essential first step to improve the outcome of teaching and learning in America's classrooms. According to reports, 48 states collaborated to write the common standards in math and reading, coordinated by the governors' group, with only Texas and Alaska refusing to participate.

Parents Need to Stay Informed As Stakeholders

Parents can stay informed by reading the reports published each year in June by the National Center for Education Statistics, referenced earlier in this chapter. As stakeholders they can voice their opinion by contacting their state congressmen and senators, as well as representatives in Washington, D.C.

Exploring Curriculum Communities and the Barriers to Change

Figure 2.1 shows the internal and external influences that add to the complexity of the curriculum development process. The influence is political from the top down, and teachers remain at the bottom, teaching from textbooks that match the state and national tests.

The walls between the communities are barriers to the 50 million K–12 students in approximately 15,000 school districts, encompassing 91,000 schools (Figure 2.2). The standards necessary to integrate nano, bio, cogno, and info science into K–12 curriculum is only an advisory function without a national curriculum. It will remain difficult to stay current with any scientific or

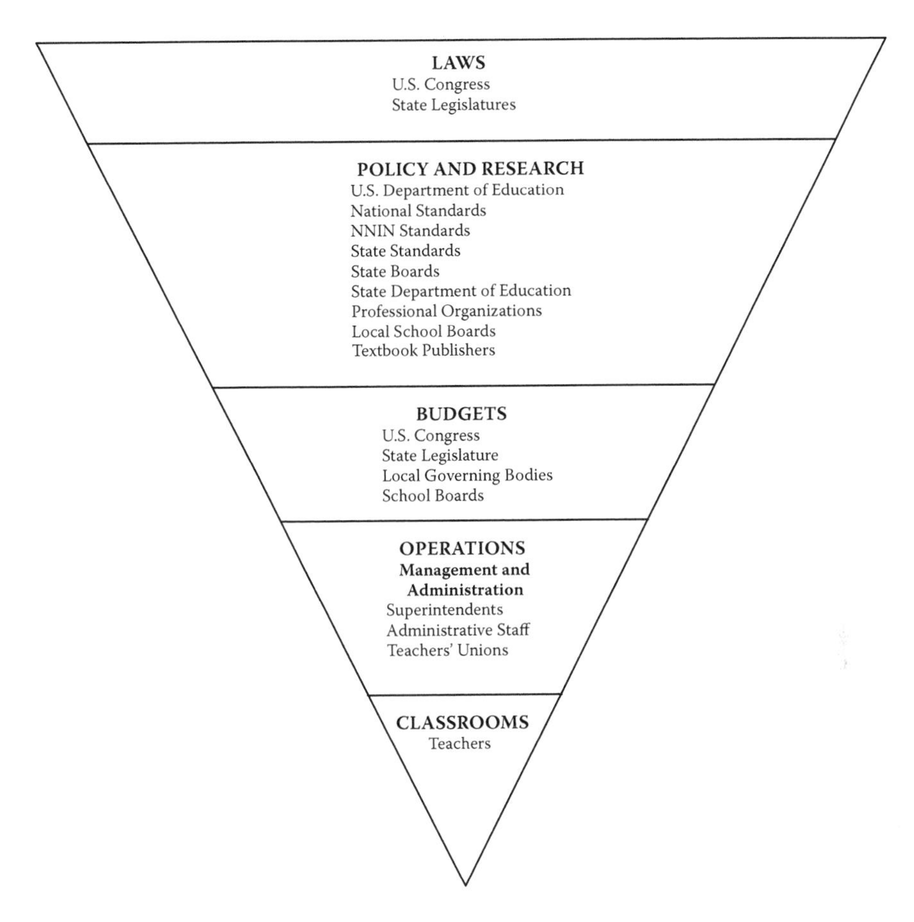

FIGURE 2.1
The curriculum communities of stakeholders.

technological advances because revisions in curriculum standards normally take five to ten years in development. Due to the laws regarding changes in educational standards, our only hope is to continue developing new curriculum that teachers may adopt in the classrooms, regardless of inclusion in textbooks or assessments.

Even though current policy uses a uniform measure of accomplishment through standardized testing, we need to develop more effective measures based on cognitive development and individual learning differences. This will also be very slow because the evaluation of new approaches typically requires a generation to see any impact. It has been difficult to identify the relationship or successful applications between these complex system approaches to address any of the key challenges of the education system. It may be fruitful to look at simpler solutions to help teachers understand where nanoscience might fit within their current syllabus.

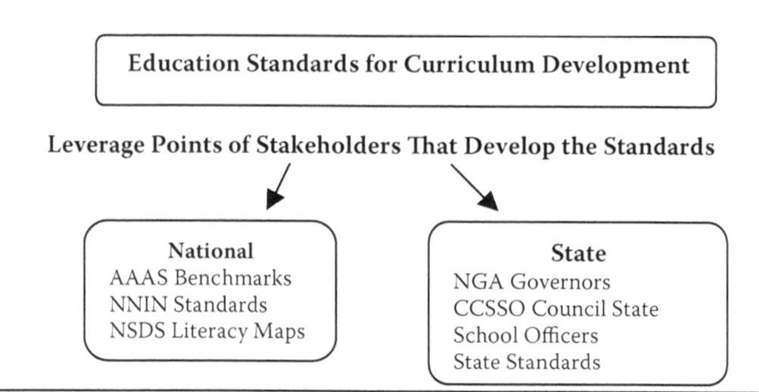

State School Boards Make the Final Decisions on Content Submitted by the Local School Boards. Textbook publishers lean towards pleasing the largest states that buy the most textbooks. This practice compromises the choices for the other 47-48 states that must choose between California or Texas textbooks.

15,000 School Districts — 49 Million Students in the U.S.
Locally taught curriculum involves rote memorization of facts published in the textbooks that will be included on the legally mandated Federal and State Assessment Testing for School Districts to receive funding. These laws "from the top down" are detrimental for teachers that would like to expand their syllabus beyond the required testing matrix with creative, interactive instructional materials that stimulate a passion for learning in grades K–12.

FIGURE 2.2
Relationships between national and state standards and local districts.

NNI-Funded University Outreach Programs Could Develop Syllabus for Textbooks

Nanoscience curriculum resources for K–12 were not developed until the National Nanotechnology Initiative (NNI) funding for Nano Centers at the universities included them in their funding as mandatory outreach programs. Even though the resources are available, teachers do not know where they fit in their teaching syllabus.

A practical solution would be for educators to introduce this scale of science to students as a new size of nature that was "too small to see" before advances in microscopy, rather than a new topic. It would then be feasible to enlist professors already developing curriculum for the university outreach programs to define where it fits by grade level. Reviewing current textbooks by state or region and aligning the interactive, visual elements developed as drop-in materials would help the teachers match the lessons to their syllabus.

The UVA Virtual Nano Lab[23] Web site at the University of Virginia, funded by NSF as the first online nano lab to develop specific lessons for K–12 students, adds the hands-on laboratory experience. The virtual nano lab has instructional materials about advances in microscopy and educational instruction for teachers on using the lab in classrooms. Adding hands-on lab experiments developed to match the textbook syllabus creates a powerful introduction. Providing a CD teachers' guide, attached to new textbooks, would alert teachers that new material is available on the Internet as a supplement. Textbook publishers are already providing CDs for teachers referring to Internet sites, so this would be a familiar and accepted approach.

Resources Developed As Outreach Are Not Guaranteed to Reach the Schools

As stated earlier, Section III of the book provides links to all of the new materials developed by the universities under the National Nanotechnology Initiative (NNI) funding. As outreach programs they are available for teachers to include as part of their science syllabus.

There are many reasons for introducing nanoscience into the teaching syllabus, even though only 17% of high school students are working at proficiency levels. Unless these changes are made in how we educate our students, with challenging up-to-date information, the number of graduate students in science or engineering-related technologies will not increase in the next decade. Parents and teachers can make a significant difference to support the changes necessary by attending local school board meetings, as both are important components of the communities of stakeholders.

Effective Collaboration Skills Are Necessary for Global Citizens[24]

As educators struggle with adopting integrated subject material, communities, states, and countries are facing new cross-border collaborative situations in economic development.[25] This also affects education as nano-enabled technology moves into the commercialization stage. The current collaboration between universities, colleges, and nano-enabled companies to define workforce education of technicians has started the process. Clusters are forming in many states as patent licensing from university research enters the marketplace. The general public has not been aware of these advances which require a very knowledgeable workforce. Students are also unaware of the demands they will face in a global market that will require not only an expanded knowledge base in the science fields, but also communication and collaborative skills. At the top national level of stakeholders, it is understood that the holistic approach to education is imperative for students to succeed in this century. What is not understood is how to implement this approach in our schools. To start the process, I have included the following case study of a highly successful role model. The potential and possibilities for global success are limitless, as defined by her example.

A More Holistic and Global Approach to Higher Science Education Is Needed in the Twenty-First Century

Case Study:

Dr. Anita Goel, M.D., Ph.D., a twenty-first century leader, physicist, physician-scientist, medical doctor, inventor, entrepreneur, humanitarian, and visionary pioneer—a role model extraordinaire—is creating a powerful difference in the world.

Dr. Goel provides the quintessential role model for the next generation of scientific leaders who will need to expand their vision of education to be more holistic and the scope and impact of their work to be more global in the twenty-first century. According to Dr. Goel, nanotechnology stems from the ability to probe and control matter at increasingly finer scales. This ability to control nature at very small scales has profound implications on a wide range of fields and industries, ranging from nanomaterials and clean energy to nanomedicine. To create and incubate breakthrough insights and disruptive technologies that emerge from the convergence of physics, nanotech, and biomedicine Dr. Goel founded Nanobiosym®[26] in 2004 in Cambridge, Massachusetts. She has pioneered the scientific field of precision-controlled nanomotors or molecular machines that read/write information into DNA. She and her group at Nanobiosym are also working on harnessing these nanomachines for various applications including novel nanomedical diagnostics, nanomanufacturing, and more advanced applications in nanoscale energy transduction and biocomputation.

Building new bridges between academia and industry and the broader geopolitical impact, Nanobiosym (NBS) works symbiotically with its commercial partner Nanobiosym Diagnostics (NBSDx) to discover, develop, and commercialize breakthrough technologies at this interface. NBSDx is currently developing low-cost nanobiotechnology platforms like the Gene-RADAR® to address global healthcare needs. The Gene-RADAR is one of Dr. Goel's inventions: Envision mobile diagnostic devices which can analyze a drop of blood or saliva and quickly identify infectious diseases and pathogens at the point of incidence.

Dr. Goel's pioneering contributions to this interface over the past 15 years have been recognized globally by several prestigious honors and awards. Her work at Nanobiosym has been rewarded by multiple awards and phases of funding from the United States Department of Defense agencies including Defense Advanced Research Projects Agency (DARPA), Air Force Office of Scientific Research (AFOSR) and U.S. Department of Energy (DOE), and U.S. Defense Threat Reduction Agency (DTRA). Dr. Goel and Nanobiosym, under her leadership, have also built a world-class team of advisors, sponsors, and strategic collaborators to realize the full potential for Gene-RADAR® in both developed and emerging world markets.

A Harvard-MIT trained physicist and physician, Dr. Goel was named in 2005 as one of the world's "top 35 science and technology innovators under the age of 35" by MIT's *Technology Review Magazine* and in 2006 received the Global Indus Technovator Award from MIT, an honor recognizing the contributions of top 10 leaders working at the forefront of science, technology, and entrepreneurship. Dr. Goel holds both a Ph.D. in physics from Harvard University and an M.D. from the Harvard-MIT Joint Division of Health Sciences and Technology (HST) and a B.S. in physics with honors and distinction from Stanford University. Dr. Goel is also a member of the Board of Overseers of the Boston Museum of Science and a charter member of TiE (The Indus Entrepreneurs, a global organization of successful entrepreneurs engaged in the cycle of wealth creation and giving back to society). Dr. Goel is a fellow of the World Technology Network, a fellow-at-large of the Santa Fe Institute, an Associate of the Harvard Physics Department, and an adjunct professor of the BEYOND Institute for Fundamental Concepts in Science and Arizona State University. She also serves on the National Board of the Museum of Science and Industry, on the Nanotechnology Advisory Board of Lockheed Martin Corporation, and on the International Advisory Board of the Victoria Institute of Science and Technology.

Nanobiosym Global Impact: An Innovative Public–Private Partnership with India

Dr. Goel launched the Nanobiosym Global Initiative to build innovative partnerships with academic, commercial, and global thought leaders, NGOs, industries, governments, and global organizations who can help bring disruptive technologies like Gene-RADAR to sustainably address some of the greatest unmet needs in both the developing and developed worlds.

Dr. Goel also serves on the board of trustees and scientific advisory board of India-Nano, an organization devoted to bridging breakthrough advances in nanotechnology with the burgeoning Indian hi-tech sector. While at Stanford, Dr. Goel envisioned building new bridges between the world's two largest democracies: the United States and India. Inspired by this vision, Dr. Goel founded and spearheaded SETU (Sanskrit for "bridge"), an international conference and think tank that brought together world leaders from academic, business, political, and humanitarian arenas at Stanford University.

As a brilliant scientist, successful entrepreneur, and global visionary, she has conceived of building multipurpose nanotechnology innovation parks in India and other parts of the developing world. The innovative public–private partnership she spearheaded, for example, between Nanobiosym and the State of Gujarat will maximize the benefits of the park for both the state and the country. The park is expected to generate billions in foreign direct investments and international commerce and create thousands of new jobs in cutting edge industries such as nanotechnology, biotechnology, high-tech manufacturing and medical tourism. By doing some of the large scale nanomanufacturing in India, Dr. Goel hopes to make the Gene-RADAR systems more affordable to address unmet health care needs of people in the developing world, with a special focus on improving health care delivery to people at the bottom of the pyramid. Her long term vision is for the Nanobiosym Technology Park in India to serve as a global hub to address unmet needs in India and other developing countries by creating a local ecosystem engaging the people and communities to improve their lives.

A parallel project: the Nanobiosym Innovation Knowledge Ecosystem enables exchanges with global scientific, technological, business, and development communities for sustainable and scalable solutions to some of the world's most pressing problems. Thus, building a holistic ecosystem to create scientific innovation at the intersection of conventional disciplines and incubating emerging technologies with potential to impact several industries, engages and uplifts the socioeconomic strata.

Understanding the Stages of Commercialization for Nanotechnology

The first phase of growth was in tools and metrology for research and development. The second phase was powders, particles, and composite materials that altered and improved products already in the marketplace such as golf clubs, tennis rackets, bicycles, clothes, building materials, composites for aerospace, etc. The third phase was new devices that are smaller and faster such as the 32-nm chips developed by Intel and the increased memory storage and faster performance at both Intel and IBM. The information sector includes advances in robotics and artificial intelligence, while the cogno sector has brain mapping projects. The nanobio field is working on targeted drug delivery of nanomedicines, cancer research, alterations of DNA, lab-on-a-chip projects for the military and space travel, and hand-held devices that detect germ warfare and perform diagnostics for disease. The possibilities for success increase as we move to phase four of complex systems in manufacturing products that have added life and value in the marketplace. According to former Undersecretary of Commerce Phillip Bond, the "wealth and security of nations depends on who can commercialize nanotechnology."

Projections in the Marketplace

The National Science Foundation (NSF), the National Nanotechnology Initiative (NNI), the U.S. Department of Education, the U.S. Department of Labor, and the U.S. Department of Commerce (DOC), all understand that the new technologies resulting from nanoscience research and development are expected to contribute upwards of 15% (*est.* $2.5 trillion) to the global GDP by 2015. NSF also estimated that this exponential growth of the sector will require two million trained nanotechnologists or technicians by 2015. This prediction draws attention to the immediate need for students to obtain practical, hands-on experience at the nanoscale level of science and engineering. The success of our future will depend on this type of education and training—"from the top-down" at our institutions of higher learning, and "from the bottom up" with our K–12 system.

New Data Shows Nanotechnology-Related Activities in Every U.S. State

"The rapid growth in nanotechnology activity across the United States illustrates the impact of continued and significant investments in nanoscience and nanoengineering by the federal government and private sector," said PEN Director David Rejeski. "There is now not a single state without organizations involved in this cutting-edge field."

Data released by the Project on Emerging Nanotechnologies (PEN)[27] highlights more than 1,200 companies, universities, government laboratories, and other organizations across all 50 U.S. states and in the District of Columbia that are involved in nanotechnology research, development, and commercialization. Their projects states that the number is up 50% from the 800 organizations identified just two years ago. As we progress through the many phases of commercialization these clusters will continue to expand at a rapid pace, adding to the diverse need for technicians trained in nanotechnologies.

PEN's interactive map[28] displaying the growing "NanoMetro" landscape, is powered by Google Maps®. It features an accompanying analysis that ranks cities and states by numbers of companies, academic and government research centers, and organizations and technology focus by sector.

Nanotechnology Map Highlights

The top four states overall (each with over 75 entries) are California, Massachusetts, New York, and Texas. These states have retained their lead since the first analysis was released in 2007.

References*

1. Source: National Center for Education Statistics, Program for International Student Assessment (PISA) conducted by the Organization for Economic Co-Operation and Development (OECD)
2. http://nces.ed.gov/pubsearch
3. "Is America Falling Off the Flat Earth?" The National Academies, Norman Augustine, Chair, Rising Above the Gathering Storm Committee. ISBN 0-309-11224-9 2007
4. Source: *Rising Above the Gathering Storm*, National Academies Press, 2007
5. http://www2.ed.gov/about/overview/fed/role.html
6. Rice University–David Ruth, http://www.media.rice.edu/media/NewsBot.asp?MODE=VIEW&ID=12907
7. http://cnx.org/content/m19435/latest/

* All links active as of August 2010.

8. The Opportunity Equation: Transforming Mathematics and Science Education for Citizenship and the Global Economy, Carnegie Corporation of New York and Institute for Advanced Study, 2009, www.OpportunityEquation.org .
9. Institute of Education Science, http://ies.ed.gov/aboutus/
10. http://ies.ed.gov/director/board/
11. http://nces.ed.gov/nationsreportcard/
12. http://eric.ed.gov/
13. http://ies.ed.gov/ncee/edlabs/
14. http://ies.ed.gov/ncer/randd/
15. http://ies.ed.gov/ncee/wwc/
16. http://ies.ed.gov/ncee/wwc/references/library/
17. NSF PR 02 87, October 23, 2002, http://nsf.gov/od/lpa/news/02/pr0287.htm
18. http://www2.ed.gov/about/overview/fed/role.html
19. http://www2.ed.gov/about/overview/budget/budget11/index.html
20. National Center for Education Statistics, http://nces.ed.gov/programs/coe/2009/analysis/index.asp
21. National Center for Education Statistics, http://nces.ed.gov/timss/table07_4.asp
22. D.J. Webber, "School Districts Democracy: School Board Voting and School Performance," *Politics & Policy*, 2010, 38, 81-95.
23. UVA Virtual Lab Website, http://www.virlab.virginia.edu
24. The Opportunity Equation: Transforming Mathematics and Science Education for Citizenship and the Global Economy, Carnegie Corporation of New York and Institute for Advanced Study, 2009, www.OpportunityEquation.org .
25. http://www.e-nc.org/pdf/Creating_Wealth_Cross_Border_Report.pdf
26. www.nanobiosym.com
27. www.nanotechproject.org/inventories/consumer
28. http://www.nanotechproject.org/inventories/map/

3

Students Are Shifting the Paradigm

> When a distinguished but elderly scientist states that something is possible, he is almost certainly right. When he states that something is impossible, he is very probably wrong.
>
> **Arthur C. Clarke (1917–2009), Clarke's first law**

Students Are Making a Difference in the Classrooms and the Workplace

Since our schools have been hampered with systemic issues and unable to easily make the necessary changes, some students are taking back control of their education. Some are choosing to become self-educated, and many are motivated to request choices in the corporate world that suit their needs. They do not just accept the established norm in the job market. They prefer opportunities for lifelong learning experiences and challenges, rather than traditional benefits, stock options, and long hours of internships. This generation was raised in the digital information age of multitasking and does not respond to boring repetitious work. They have established different values and preferences in the workplace, demanding open-ended educational opportunities, flexible time schedules, and remote location working environments.

This new generation of workers have been classified as the Millennials, and were born between 1978 and 2000. As they enter the workforce they become one of four generations that must learn to coexist, according to an article titled "Scenes from the Culture Clash."[1]

The article also describes the other generations as: Traditionalists (born before 1945), the Boomers (1946–1964), Generation X (1965–1977), and the Millennials (1978–2000). As you would expect, it is very difficult for managers to minimize the friction and maximize the assets of these four distinct sets of work styles. The Millennials are interested in personalizing their careers with learning opportunities, and they will dominate the workforce for the next 70 years.

How Did We Miss Preparing Management for This Talented Generation?

Experts ignored the fact that this generation was immersed in technology. Computers, video games, e-mail, the Internet, and cell phones dominated their childhood. They have no fear of learning or achieving their goals. They are truly a generation of self-achievers. They are also the first generation that desires to continue achieving through "Lifelong Learning." Experts have discussed the potential of continued learning, but have not established a protocol within our educational structure.

How Do These Young Professionals Fit into Our Establishment Now?

Young lawyers were once willing to work 100-hour weeks for many years for the chance to become a partner. That is now history. Law school graduates from this generation want work–life balance, flexible schedules, and philanthropic work. This generation is affecting the entire spectrum of the workforce including financial firms such as Deloitte and Touche USA. The firm has been testing a program in New York for new employees to work remotely. The old way of camping out onsite at a client's company was not acceptable to the new Millennial workers. The test worked out so well, they expanded the program nationally.

Marriott International had to change their style of training to match the Millennials' rapid-fire style of information consumption. They are now developing "bite-size edutainment" training podcasts. Workers can download them to their cell phones, laptops, and iPods as needed. Podcasts are also being developed in universities to replace the traditional classroom lectures.

The new workers insist on relationships with top management and want to be heard when they speak. They value respect and prefer to build relationships in the workplace, based not on titles or hierarchy, but respect for ideas and human interactions. They are not asking for signing bonuses or stock options. They just want to be heard, and we might just learn something from them if we take the time to listen, which may lead to real ethics and values in the corporate workplace.

So How Do These Generational Changes Fit into a Collaborative Advantage for Education?

Participation: Government officials and politicians could invite input from this new generation of workers.

Mentoring: Invitations to join national "think tanks" at an early age would prepare them for leadership roles for political offices in the national arena.

Collaborate: Enlist these young talented people to collaborate with cognitive experts to expand their research of how students learn to include their learning styles.

Observe: Cognitive researchers could observe how young students are helping teachers learn to master technology and sharing knowledge with students who lack technology at home.

Share: Use team teaching to change methods of teach/learn knowledge sharing in the classrooms. Interweave subject material from textbooks into multiple topics through storytelling and role-playing situations. This identifies difficult concepts of math and science as scenarios that match aspects of student's lives. Art and music support storytelling, while reading and writing skills improve when subjects are interwoven naturally in early grades.

Teaching Nanotechnology in Grades 1 through 6 in Singapore Was Initiated by an 11-Year-Old Girl

A perfect example of interwoven teach/learn activity was initiated in Singapore, introducing nanoscience and nanotechnology to grades 1 to 6 in January 2005. Balestier Hill Primary School developed a nanotechnology program for all students in primary grades 1 to 6, wherein the school set up a $25,000 air-conditioned nano lab for "hands-on" experiential lessons. Associate Professor Belal Baaquie initiated the project based on a request by his daughter Tazkiah, age 11. He knew that nanotechnology was an emerging area in science and technology and felt strongly that students should be exposed to it from a very young age, when they are open to new ideas.

In December 2004, Professor Baaquie gave a talk on nanoscience to the school principal, Dr. Irene Ho, and the teachers. He then organized a visit for them to the university labs, which convinced Dr. Ho, and she swung into action. She quickly submitted a proposal to the Ministry of Education and was given a $15,000 grant from the School Innovation Fund, with another $10,000 from the School Cluster Fund.

A month later (not years in the decision-making process), the lab was ready. All the children go to the lab two or three times a week. Lessons are made fun and simple, especially for the younger ones. Under the teacher's supervision, the children in primary grades 1 and 2 are allowed to look through the microscopes. They are then encouraged to talk or write stories about their experience to familiarize themselves with a science lab and the equipment.

They are also improving their speaking, reading, and composition skills. Children in primary grades 3 and 4 are more advanced and use golf balls and Lego sets to learn how to construct models of atomic structures.

Things get a little more in-depth for students in primary grades 5 and 6, as they are permitted to use the microscopes independently. To learn scientific protocol, they examine a strand of hair, learning to first observe, then record their findings on worksheets, which gives them an actual research lab experience. The students in primary grade 6 must develop a project involving nanotechnology, in addition to the lab research experiences.

It was decided that the students would not be tested as the scientific experience becomes part of their syllabus by being integrated with other subjects. Because an art teacher can book the lab and ask her pupils to draw what they saw under the microscope, while a language teacher can ask them to write a story, this early experience of team teaching creates a space for everyone to learn together. The lab was designed to look like a creative high-tech play room to stimulate an experience of fun while learning that is nonthreatening to the children. In one corner stand two eye-catching rectangular floor lamps and the walls of the lab—both inside and out—are covered with bright, bold wallpaper, featuring graphics of atoms and molecules. This extends the idea that art and science are both creative aspects of nature, but nano- and microscales are so small that tools are necessary. The lab was developed with eight electron microscopes—×1600 resolution models. Students can see objects the size of a micron, which is about the size of a dust particle. Each microscope costs about $3,000, and it was decided that this was economical enough for a beginning introduction. The interactive corner has Lego sets and golf balls for the model-making assignments, and display cabinets feature their finished projects and artwork.

What Can We Learn from This Example of Teaching in Singapore?

Building labs in 91,000 elementary schools is too expensive, but teachers could explore the University of Virginia (UVA) virtual nano labs,[2] which were funded by the national Science Foundation (NSF), on the Internet in their classrooms. The project provides visual teacher training sections, lesson plans, and experiments to introduce nanoscale teachings and experiments in classrooms.

Systems Thinking for Solutions in Education

Dr. Fritjof Capra, Ph.D., physicist and systems theorist, and winner of the "Leondardo DaVinci Society for the Study of Thinking Award–2007,"[3] is also a founding director of the Center for Ecoliteracy[4] in Berkeley, California. The center promotes ecology and systems thinking in primary and secondary education, based on one of his earlier books titled *"The Web of Life."*[5] Capra took a giant step in setting forth a new scientific language to describe the interrelationships and interdependencies of psychological, biological, physical, social, and cultural phenomena in the web of life. The Center has programs to assist K–12 schools in learning how to Green the Curriculum. The following example is a story of change initiated by a high school student that wanted his school to Go Green.

Another Student Initiative That Led to the Greening of a K–12 Curriculum

Through determination, a student of Head-Royce School, a K–12 independent school in Oakland, California, convinced the principal to investigate a program to green its entire curriculum.

In 2006, Alejo Kraus-Polk, a 15-year-old sophomore, needed permission from Principal Paul Chapman to invite the executive director of the Berkeley-based Green Schools Initiative to Head-Royce, because he wanted his school to go green. Shortly after, Yaeir Heber, a junior, was elected student council president on a platform emphasizing environmental action. Chapman was aware of a growing student crusade for the environment and considered Kraus-Polk and his mission a worthwhile endeavor, approving the request and attending the talk.

After the presentation, Chapman decided the most important first step was to signal commitment to the students from the top down, and persuaded the board to approve a green mission based on these four goals:

- Create a healthy environment.
- Use resources in a sustainable way.
- Develop an educational program.
- Pursue a nutritional health program.

These defined goals were critical to the success of the project. Greening the school did not mean just focusing on the normal activities of composting and recycling, or even greening the building. Discussions were initiated to develop a green curriculum and what that would mean in the classroom.

The project immediately took hold and cemented the desire to accomplish these goals with everyone on board. Students led the activities and formed a green council of 12 voting members, again, mostly students.

The Green Graduate

What would a green graduate look like? The principal started searching for answers and soon found the book titled *Ecological Literacy: Educating*

Our Children for a Sustainable World,[6] developed and edited by the Center for Ecoliteracy.[7] As Chapman read it, he found that one essay provided the framework that became crystal clear. He started to understand that a sustainable society needed to embrace a new way of seeing the world. Capra called it systems thinking, and it emphasized relationships, connectedness, and context in any type of system, including schools.

Capra explained them as perceptual shifts, and there were eight important concepts that would describe patterns and processes in nature that sustain life. Theses are networks, nested systems, interdependence, diversity, cycles, flows, development, and dynamic balance. He called them principles of ecology, principles of sustainability, principles of community, and curriculum that would teach children these fundamental facts of life.

Chapman was convinced, and meetings with all departments confirmed they would apply the Center for Ecoliteracy's principles throughout the K–12 curriculum. Again, as we saw in the last example in Singapore, this decision did not take the normal time frame of a year to approve; it happened in the first meeting, and it became a grassroots endeavor from the bottom up.

Integrated Teaching of Subjects Promotes Sustainability

In the process, they proved that sustainability could be integrated. It started with science, but also included math, literature, history, ethics, world languages, and art. This integration instigates systems thinking, and soon everyone was amazed, while recognizing they had barely just begun.

Capra Reveals Leonardo's Artistic Approach to Scientific Knowledge

Taking integration of subjects a step further brings us to Capra's recent book, *The Science of Leonardo*.[8] In this book, Fritjof Capra reveals Leonardo da Vinci's artistic approach to scientific knowledge along with his organic and ecological worldview. The principles used in his designs for rebuilding Milan, Italy, are still used by city planners today. *The Science of Leonardo* is the first book to present a coherent account of the scientific achievements of da Vinci, the great genius of the Renaissance. He evaluates them from the perspective of twenty-first century scientific and philosophical thought. Its central thesis is that Leonardo's science is a science of living forms, of quality. This can be seen as a distant forerunner of today's complexity and systems theories, linking it to his brilliant synthesis in the *Web of Life*.

Da Vinci's life work is a science that honors and respects the unity of life, recognizes the fundamental interdependence of all natural phenomena, and reconnects us with the living Earth. Da Vinci's science is thus highly relevant to our time. The book has been published in seven editions in five languages.

Through this meticulously researched book, da Vinci emerges as the unacknowledged "father of modern science." He is the perfect role model for combining art and science in the classrooms, encouraging young children and older students to "connect the dots in the study of nature."

The book also gives us a clear and profound look at the life and complexities of an artist from the Renaissance era. He had a unique ability to understand the science of nature and technology through his art. This is a perfect example of a road seldom taken by Western science. It is also a perfect resource book for art and science teachers to glimpse the mind of an artist who understood the nonreductive science of systems in nature through his art. During his lifetime he created over 6000 pages in notebooks and used a backwards writing code to protect his work. These are now being analyzed by experts to better understand the genius of his mind.

Introducing Nanoscience through Art

In order to encourage teachers to explore this unique area of collaborative team teaching, Cris Orfescu, an artist and scientist, joined me in developing an online exhibit for the nanoscale artwork of K–12 students. The purpose of this global program is to stimulate creativity while exploring nature, to expand the visionary imaginations of our children, to promote a new paradigm unifying the art-science-technology intersections at the nanoscale.

NanoArt is a new discipline to combine art with science and create paintings or sculptures based on visual scans from the nanoscale of 1 to 100 nanometers. Art is the perfect media for this first introduction to the visual scans. The strange surface topography at this tiny size of nature yields familiar shapes that are easily recognized in the finished artworks. All compositions are grouped on the Internet by age/grade level for the viewers.

The oil pastel painting from the nanoflower scan (Figure 3.1) serves as an example of an image reflected in the patterns and created as an example for teachers (Figure 3.2).

The *NanoArt for Kids* program allows teachers to explore the outsource materials provided by universities. Teachers can also order free scans for their classrooms as part of the project.

You can download the PDF file of the recommended modules and the form for submitting artwork.[9]

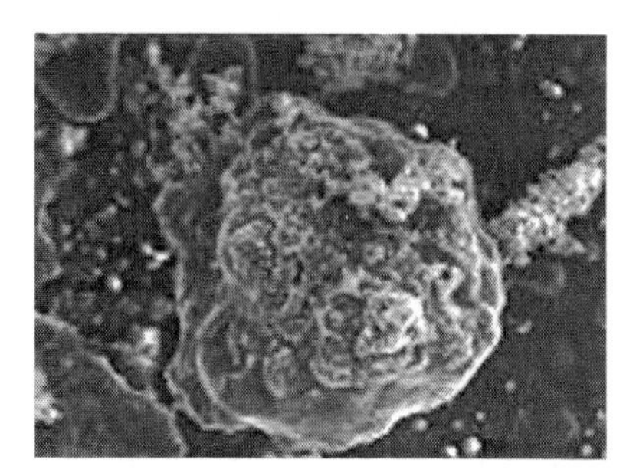

FIGURE 3.1
Nanoflower scan.

FIGURE 3.2
"Nano Wisdom" oil pastel painting.

Science, Art, and Writing (SAW): Breaking Down the Barriers between Art and Science

Anne Osbourn, from the John Innes Centre, Norwich, United Kingdom, contributed a paper[10] that explains the importance of breaking down the barriers between art and science. The following excerpts may help to clarify the importance of this integration of subjects.

> The path to specialization of knowledge starts early. By the time children leave primary school, they have already been taught to view subjects like biology, art, and social studies as unrelated disciplines rather than as interlocking pieces that together lay the foundation for a deeper understanding of the world. The divisions between science, the arts, and the humanities are reinforced in high school, where each subject is taught by a different instructor under pressure to "teach to the test," a practice that further isolates subjects, stifles inquisitiveness, and quells creativity. By the time we become specialists as adults, our ability to recognize connections between disciplines tends to diminish even further—often at a price.
>
> While high school students and adults often feel constrained by mental barriers, elementary school children have not yet been programmed into compartmentalized ways of thinking and have fewer inhibitions. They explore the world around them through personal adventure and discovery, inquiring, speculating, exploring, experimenting, and risk taking.
>
> Children need these skills to realize their full potential and creativity as they grow into adulthood; scientists need them to break new ground. Without the ability or confidence to take risks as adults, whatever area we choose to specialize in, we are unlikely to make important contributions to our fields.
>
> We all try to make sense of the world around us in our own ways, and although scientists and artists clearly approach their occupations in different ways, both depend on the ability to define a problem, note

detail, inquire, and extract the essence of the problem in hand. Both require a combination of creativity and technical competence. The core of the scientific process entails critical thinking, the generation and testing of hypotheses, and the rigorous interpretation of data/observations, which in turn leads to further speculation, prediction, and testing. The general perception of artists seems to be that they engage in far more unrestrained interpretations of natural observations in order to understand the world around them, although I personally would be reluctant to suggest that these specialists are incapable of critical thought, experimental design, rigorous analysis, and iterative progressive development. These issues aside, both groups bring their own preconceived ideas, skill-sets, and perspectives to the problems they take on. Ultimately, they aim to pinpoint the "best truth possible" available to them at a given moment in time and to communicate this understanding clearly and succinctly to others for appraisal. As more perspectives are assimilated into these individual understandings, we can build up a composite, more refined and durable understanding.

Sam Mugford (top left) and Melissa Dokarry (top right) showing 7- to 9-year olds at Martham Primary School how to extract and analyze pigments from plants.

One way to integrate science into the lives of students takes advantage of the natural curiosity of young children and the power of visual images to engage that curiosity to investigate science and the world. Stunning scientific images—particularly those that show something in an unusual way—act like magnets, attracting people of all ages and disciplines with their intriguing, nonthreatening representations of natural phenomena. They awaken curiosity—a hunger to learn more. By using images from science as a starting point for scientific experimentation, art, and creative writing, the Science, Art and Writing (SAW) initiative[11] breaks down barriers between science and the arts. Each SAW project has a scientific theme, supported by a collection of carefully selected visually striking scientific images. With this approach, children realize that science and the arts are interconnected—and they discover new and exciting ways of looking at the world.[12]

Students Are Digital Natives Who Are Now Teaching the Teachers

A survey of teachers and instructors at the high school and postsecondary levels has found that students who excel in the use of information and communications technology (ICT) are driving change in classroom instruction. The study was carried out by Certiport Inc.,[13] a provider of technology training, certification, and assessment solutions, and the Education Development Center Inc. (EDC), an international nonprofit organization that researches and implements best practices in health and learning in 50 countries. The survey of 444 teachers and instructors was conducted in 382 Certiport testing centers over a seven-day period. Power Users, as defined by EDC, are the savviest of the "digital natives," a demographic of 10- to 15-year-old students who have grown up with digital technology as a part of their everyday lives.

According to EDC, these students have technical acumen beyond any previous generation. They are characterized by their ability to "leverage the Internet to the highest degree conceivable" and are energized by technology well past the point of most digital "immigrants"—that is, older learners forced to adapt from the analog age. This group is in tune with skills that are needed for success in the twenty-first century, exhibiting many of the collaborative learning, analytical thinking, and problem solving interests that are sought by today's employers.

Power Users exhibit engineer-level thinking that we do not normally expect students to have until they enter postsecondary engineering programs. Among the survey findings: 69% of respondents believe Power Users influence what is being taught in the classroom, and 66% said they influence teaching methods. According to the survey, 48% of respondents said Power Users exhibit helpful behavior, and 55% said these students facilitate the learning of other students. Teachers, meanwhile, are pairing these students with other, less technically advanced classmates in hopes that they will assume more of a leadership role and are encouraging them to share their breadth of knowledge with their peers. The study also found that more than four in five teachers (84%) believe Power Users have positively influenced their own learning and knowledge of ICT. A synopsis of the report is available at the EDC Web site, along with other materials related to the four-year Power User study.[14] This phenomenon sets the stage for team teaching as a component in the future of education globally. As schools consider technology versus textbooks due to state budget deficits, these Power User students ensure a successful transition in the classrooms.

Study Shows Four-Year-Old Preschool Students Think Like Scientists

Scientific studies show that preschool children think like scientists. The children were convinced that perplexing and unpredictable events can be explained, according to an MIT brain researcher. The way kids play and explore suggests they believe cause-and-effect relationships are governed by fundamental laws rather than by mysterious forces. Laura E. Schulz, assistant professor of cognitive science and co-author of the study "God Does Not Play Dice: Causal Determinism and Preschoolers' Causal Inferences,"[15] and her colleague, Jessica Sommerville of the University of Washington, tested 144 preschoolers.

The purpose was to look at whether children believe causes always produce effects. If a child believes causes produce effects deterministically, then whenever causes appear to work only some of the time, children should think some necessary cause is missing or an inhibitory cause is present.

In one part of the study, the experimenters showed children that a switch made a toy with a metal ring light up. Half the children saw the switch work all the time; half saw that the switch only lit the ring toy some of the time. The experimenters also showed the children that removing the ring stopped the toy from lighting up.

The experimenters kept the switch, gave the toy to the children and asked the children to stop the toy from lighting up. If the switch always worked, children removed the ring. If the switch only worked some of the time, children could have removed the ring but they did not—they assumed that the experimenter had some additional sneaky way of stopping the effect.

Children did something completely new: they picked up an object that had been hidden in the experimenter's hand (a squeezable keychain flashlight) and used that to try to stop the toy. That is, the children did not just accept that the switch might work only some of the time. They looked for an explanation.

Conclusion

This is the first study that looks at how probabilistic evidence affects children's reasoning abilities concerning unobserved causes. This research suggests that preschoolers actually have quite abstract beliefs about causal relationships. Most schools in the United States do not introduce science as a subject until grades 3 or 4, missing the opportunities to stimulate children who are naturally inquisitive and open.

More Nursery School Children Going Online

Before they can even read, almost one in four children in nursery school are learning a skill that some adults have yet to master: using the Internet. Some 23% of children in nursery school—kids age 3, 4, or 5—have gone online, according to the U.S. Department of Education. By kindergarten, 32% have used the Internet, typically under adult supervision. The numbers underscore a trend in which the largest group of new users of the Internet are kids 2 to 5 years old. At school and home, children are viewing Web sites with interactive stories and animated lessons that teach letters, numbers, and rhymes.

In order to understand this phenomenon, we must realize that young students do not differentiate between the face-to-face world and the Internet world. They were born into the age of the Internet. They see it as part of the continuum of the way life is today.

Scholastic Inc. has a section of its Web site designed just for children who go online to read, write, and play with *Clifford the Big Red Dog.*[16] PBS Kids Play[17] Online has more than a dozen educational Web sites for preschool children, including *Sesame Street* and *Barney and Friends.* Overall computer use, too, is becoming more common among the youngest learners. Department figures show that two-thirds of nursery school children and 80% of kindergardeners have used computers.

Virtually all U.S. schools are connected to the Internet, numbering about one computer for every five students, the government reports. Many older students are often far ahead of their teachers in computer literacy, and they realize their younger siblings are gaining on them.

Teaching the Art of Game Design As a Career Path Combines Art and Computer Science

More than 100 colleges and universities in North America, up from less than a dozen five years ago, now offer some form of video game studies. The late Randy Pausch, former codirector of the Entertainment Technology Center at Carnegie Mellon University,[18] which offers a master's degree in entertainment technology, stated that gaming studies have a sneaky side: They attract students to computer science.

At the Summit on Educational Gaming, sponsored by the Federation of American Scientists,[19] a statement was made that video games have the potential to improve learning in the United States and keep the nation at the forefront of global competition. Educators and cognitive scientists joined forces with software marketers and designers to discuss the possibilities of merging digital gaming into the education of what some described as our increasingly attention-deficient society. The average teenage male spends approximately 316 hours playing video games each year. The hope is that by producing content in a medium already familiar to, and welcomed by students, more of them will be able and willing to master basic knowledge and skills.

Mike Zyda, director of the University of Southern California's GamePipe Laboratory,[20] reported that he sees video games taking on a much more active "teacher" role. GamePipe, an R&D laboratory for interactive games and their practical applications, is currently collaborating with a private company on a piece of technology that would do just that. The plan is for a noninvasive sensor that can monitor a player's brain activity in order to gauge the rate of learning, the modality of learning, and the person's emotional state. Once this information is processed, the software would tailor the game's activity to best fit the comprehension methods and speed to which the player seems to respond best.

Learning to leverage the popularity of video games to help students excel was the core purpose of a Serious Gaming Conference[21] event also held in 2005, in Washington, D.C. The Federation of American Scientists (FAS) Summit on Video Gaming wanted to demonstrate the pedagogical value of gaming technology, often viewed with skepticism by generations of educators who did not grow up in the digital age. The FAS event focused on the theory behind using video games in the school curriculum and looked at how to use gaming curricula to engage students and improve their performance.

The summit ended with two panel discussions on innovation in gaming, one of which focused on the challenges to innovation in the education and training markets.

There is nothing shocking about the use of computer gaming in classrooms, of course. The goal has been to replace the earlier "drill-and-practice" methods of interactive learning with a new generation of pedagogical tools for all educational levels and in subjects ranging from science, mathematics, and engineering to social sciences and humanities. One of the seminal programs in the field was MIT's Games-To-Teach Project.[22] Since 2001 MIT has developed more than a dozen interactive and Web-based games with names like "Replicate," "Biohazard," and "Revolution."

Some even see games and gaming technology as the key to keeping U.S. workers competitive in the world marketplace. One of those is the Digital Media Collaboratory,[23] one of several technology laboratories at the University of Texas at Austin's IC2 Institute. They work with partners from the public and private sectors to develop computer games that can be used by schools, businesses, and governments. Austin is home to several of the largest online gaming companies. The decision to start the laboratory grew out of the institute's successful use of simulations to train welfare recipients.

A pilot program was created in 1998 called EnterTech, a 45-hour training simulation that taught 44 entry-level job skills through digital role playing. The results stunned everyone. Of the 238 participants, two-thirds of the group either found work or enrolled in continuing-education programs.

Those who worked received a $1.06 average increase in salary. Bolstered by that success, the group began tailoring programs for different organizations. Versions of EnterTech have since been used in the Dallas Independent School District, the University of Texas, at-risk community schools in Waco, Texas, and adult learning centers and welfare offices throughout the state.

Despite the success of programs like EnterTech, the video-game industry has not been proactive with schools. Educational game sales make up only 7% of the software market for console games, and computer games have not generated enough sales to be ranked, according to the Entertainment Software Association.

The Digital Media Collaboratory, is a public/private partnership that received a highly competitive grant from the Texas Workforce Commission to create *Get There Texas* (GTTX). *Get There Texas* will provide an interactive, participatory Web site that links students, employers, and educators.

Built on a social networking paradigm, GTTX's online collaboration tools will connect business and education and improve the alignment of the Texas workforce pipeline for emerging technology-intensive industries. The project will enable employers and educational institutions to stimulate interest and encourage students and adults to pursue careers in new and emerging fields (including nanotechnology-trained technicians) from the programs developed at Texas State Technical College.

First Nanoscience Educational Game for K–12 Developed in the United Kingdom

Recommended as outreach for K–12 students by the College of Nanoscale Science and Engineering (CNSE) of the University at Albany, the first nanoscience educational game, titled *Nano Mission,* was created by PlayGen Inc.,[24] under the creative direction of Kam Memarzia. The game is online free for students and teachers to use in classrooms around the world. The game now has four modules, and students are posting videos of the game in play on YouTube. It is the perfect introduction to nanoscience in elementary classrooms at no cost to the school districts. PlayGen is a leading developer of serious games and simulations proven to improve performance, knowledge, and behavior. Memarzia teaches game development and conducts the Serious Game workshops for educators and game developers in the United Kingdom.

Essential Features, Content, and Pedagogical Strategies in Game Development

Nanotechnology and nanoscience are concerned with a new and unique set of emerging behaviors of matter—those that are observed at the border of quantum effects or in the 1–100 nanometer range. Since this range is relatively new, the introduction through visual learning techniques such as an interactive game platform ensures the retention of difficult material.

Integrating Science and Mathematics

Scientists generate data and use mathematics as a tool for data analysis. Yet, in our education system, students see these two subjects as separate and distinct. We have chosen to teach them separately, giving students a dichotomous view of science and mathematics. A teaching strategy that could be used to make connections between science and mathematics is visual learning. Combined with problem-based learning, the visual elements within game platform technology help the learner create a visual picture of concepts. They make connections between mathematics and science, through discovery learning experiences. This accomplishes the goal of combining real-world examples connecting math and science. Students are prompted through the game to demonstrate their understanding of the concept before they can move forward. Virtual examples and role-playing within the game keep the information interesting and challenging. Critical thinking skills are addressed within the platform of the game design.

How People Learn through a Gaming Platform

First, technology supports a real-world context for learning by using simulations built into the gaming platform. This forms the basis for project-based learning, to cover course content and fulfill certain course objectives. The simulation in the game exposes students to real-world problems to which they must find solutions. They are looking for answers that are "situation specific" rather than the "right answer."

Second, technology connects students with outside experts. Through the guidance of scientific advisers, students can have access to experts in any field globally. Students can also download documents through resource recommendations not available in school libraries.

Third, rather than talking about concepts, the gaming platform visually explains them. For example, the advanced imaging technology within a game simulation allows the players to look at the atomic structure for a new fuel source. Also, 3-D imaging allows gaming developers to include chemistry for students at the nanoscale by constructing a three-dimensional model of an atom. The periodic table can be animated, and students can identify the composition of elements visually. Modeling can also show the chemistry outcomes as a process from choices students make in combining elements for bonding. By trial and error, visually experiencing the results of a choice that will not bond in chemistry helps them remember the elements and understand why they failed.

In mathematics, simulations include graphing calculators for students to see the relationships between variables. Concept mapping helps students visualize processes and relationships for problem solving and spatial issues.

Fourth, technology provides scaffolds for problem solving, such as the online games that have adaptive physics engines. In today's rapidly changing world students need to learn much more than the knowledge written in a textbook. They need skills to examine complex situations and define solvable problems within them. They need to work with multiple sources and media. They need to become active learners and to collaborate and understand the perspectives of others. Students today need to learn how to learn; that is, they need to learn how to ask questions and identify problems.

- Investigate: multiple sources/media—a game Web site can provide links to education centers and related material.
- Create: engage actively in learning through role playing of the characters within the game structure.
- Discuss: collaborate; diverse views—since each student picks a role to play in the game, they position themselves as teams and after each session discuss their choices and collaborate online in the game community for the next part of the game.
- Reflect: learn how to learn—this aspect involves learning how to think, make critical decisions in the moment, and develop an incredible sense of self-respect for their new skills.

A gaming platform designed to create role-playing as adventures with robotics and engineering, creating real experiences in the diverse fields of science, stimulates a desire for knowledge. Roles that demand the students' comprehension of societal implications that arise based on their situational behavior and decisions, stimulate responsibility and maturity.

Fifth, technology provides collaboration between students and outside experts that helps problem solving. Through e-mail, discussion boards, and a game Web site, students have access to educators and experts who can guide them to think through problems. Also, interrelated Internet technology provides students with problem-solving experiences by developing "Inquiry Units." A model for this is currently available to both teachers and students through a Web site at the University of Illinois at Urbana-Champaign.[25] Development of an "Inquiry Page" is more than a Web site.

It is a dynamic virtual community where inquiry-based education can be discussed, resources and experiences shared, and innovative approaches explored in a collaborative environment.

The Inquiry web resource is based on John Dewey's philosophy that education begins with the curiosity of the learner. Utilizing a spiral path of inquiry by asking questions, investigating solutions, creating new knowledge to gather information, discussing discoveries and experiences, and reflecting on new-found knowledge, students who have played the game then create and develop their own virtual community.

Visual learning elements help students and teachers make connections between math and science. There are three types of connections made. First, there was a "data" connection; second, there was a "language" connection; and finally, there was a "life" connection.

Data Connection

Students have the opportunity to make connections between math and science by using mathematical formulae and concepts to analyze and draw conclusions from data they generate and experience during the game.

For example, if the students are playing a space exploration game, they will enter a section of the game specifically designed for them to plan a route for their adventure. They must vector all the different angles and then measure the angles and distance of each trajectory. The students then use vector concepts and formulas to determine the best angle for launch and liftoff to achieve the greatest distance on their fuel supply. In their reflections on this simulation, students will understand the connections they saw and experienced between mathematics and science. They will discover that the game had connections from Algebra II to physics in different ways. Due to unique attributes in virtual reality construction and the physics game engines, all aspects of the experience can be interwoven as a skill level process for student development and challenge.

Language Connection

Second, visual elements help students and teachers begin to form a common math/science language. The advantage of a common language to the student is they begin to see a common set of terms for a concept, rather than two sets, one in math and one in science. Teachers also see this as a big advantage. One math teacher stated that a common terminology would help the students make connections between the two disciplines because, although math and science have common concepts, they use different terms to describe them. One of the administrators also added that a common language would also serve to make the curriculum more cohesive. Addressing this concern by unifying the language in the game matrix establishes the process of unifying the sciences.

Life Connection

This is an important aspect of gaming as a learning platform. The simulations assure the experience of a concept as described. Game simulations can accomplish real-life experiences with virtual worlds.

Conclusion

In conclusion, developing visual learning elements is not an easy task. Technology, and access to technology, is an important component for developing visual tools. Both students and teachers see visual learning as a common link that helps all students have access to the concepts. Students, who may tune out during a lecture and may be left behind as a result, actively engage in the learning process through visual and experiential simulations of role playing in virtual games.

Technology is one of the factors fueling the changes we are seeing in the world by providing an expanded information base. No longer is information confined to physical libraries as the Internet provides access to libraries and resources worldwide.

Students, also faced with challenges associated with discovery learning, find that sometimes there might not be a right answer to a problem, and that the teacher may not have all the answers. Through visual learning experiences in a game platform, students develop shared knowledge, encouraging a sound understanding of big concepts in math and science and how they relate to their lives.

Role Playing As Experiential Learning

Along with visual learning elements that help students retain information and concepts for a learning experience, the game platform includes role playing and experimentation for hands-on virtual experiences, requiring choices to advance in game play. In terms of pedagogy, experiential learning is a process by which the experience of the learner is reflected upon, and from this emerges new insights and knowledge.

Experiential Learning Model

Most models of experiential learning are cyclical and have three basic phases: an experience or problem situation; a reflective phase within which the learner examines the experience and draws learning from that reflection; and a testing phase within which the new insights or learning, having been integrated with the learner's own conceptual framework, are applied to a new problem situation or experience.

The theoretical work done on experiential learning has established it as a method of learning which is useful to both educators and learners. This methodology helps learners to develop capacities to reflect on experience and appropriate significance through such reflection.

ESA Highlights Online Games as Key Learning Technology

A current example comes from the European Space Agency's (ESA) Technology Observatory, which recently tasked a study, *Online Game Technology for Space Education and System Analysis,* to look at potential applications of different online game-playing technologies from the simplest content-oriented games through to Massively Multiplayer Online (MMO) virtual worlds.

The study highlights a number of ways in which these technologies could benefit ESA aims: immersive environments based on these technologies could enhance collaborative working of project scientists and engineers. It was also recognized that exciting online games could prove an excellent tool for promoting space and supporting the teaching of science, technology, engineering, and mathematics. As part of the study, a video by Mindark[26] of a potential future game environment was produced, showing future human exploration of Jupiter's ice moon Europa.

Secondary school and university students are considered as the natural target audience of such exploratory learning environments, being already familiar with the interaction principles involved. But other important groups are also recognized: educators, members of the public without any previous interest in space, space professionals, parents, and current game enthusiasts. Widespread consultation concerning the design and promotion of any potential product would be required for such an initiative to become a successful educational tool. Therefore, ESA experts and representatives would need to involve parents and educators, national space agencies, and industrial contractors.

NASA MMO[27] Game "Moonbase Alpha"

NASA has been exploring games as education for the past decade and is hoping to create a very popular online gaming/educational experience that will not only entertain, but interest young people in careers in science and engineering.

The NASA Learning Technologies (LT) at Goddard Space Center has initiated the project after studying gaming environments since 2004. They have found that synthetic environments can serve as powerful "hands-on" tools for teaching a range of complex subjects. Virtual worlds with scientifically accurate simulations could permit learners to tinker with chemical reactions in living cells, practice operating and repairing expensive equipment, and experience microgravity, making it easier to grasp complex concepts and transfer this understanding quickly to practical problems. MMOs help players develop and exercise a skill set closely matching the thinking, planning, learning, and technical skills increasingly in demand by employers. These skills include strategic thinking, interpretative analysis, problem solving, plan formulation and execution, team-building and cooperation, and adaptation to rapid change.

The power of games as educational tools is rapidly gaining recognition. NASA is in a position to develop an online game that functions as a persistent, synthetic environment supporting education as a laboratory, a massive visualization tool, and collaborative workspace while simultaneously drawing users into a challenging game-play immersion.

There are concerns that a NASA space reality platform may not be very popular, because other in-space universes offer science fiction or space fantasy, with epic spaceship battles and alien encounters. Hopefully NASA and the MMO developers will strike a healthy balance between education and entertainment. Developing the game with an underlying story to keep the players interested promotes learning while having fun. Student input during development and testing phases would also be wise—because the game has to be fun and challenging from their perspective, not ours.

References*

1. D. Sacks, "Scenes from the Culture Clash," Fast Company, Jan/Feb 2006.
2. http://www.virlab.virginia.edu/VL/home.htm
3. http://www.davincithinking.org
4. http://www.ecoliteracy.org/education/sustainability.html
5. Fritjof Capra, *The Web of Life,* 1996, ISBN 0-385-47675-2.
6. David W. Orr, *Ecological Literacy: Educating Our Children for a Sustainable World,* 2005.
7. http://www.ecoliteracy.org
8. Fritjof Capra, *The Science of Leonardo,* 2008, ISBN-10-1400078830
9. http://www.tntg.org/documents/projects2009.html
10. A. Osbourn, SAW: Breaking down barriers between art and science. *PLoS Biol,* 6(8), e211, 2008. doi:10.1371/journal.pbio.0060211
11. SAW initiative can be found on the Saw Trust Web site, http://www.sawtrust.org/
12. *PLoS Biology,* August 2008, 6(8), e211-photo, doi:10.137/journal.pbio.0060211.g002, http://www.plosbiology.org
13. Certiport Inc., http://www.certiport.com
14. Education Development Center Inc., http://www.edc.org

* All links active as of August 2010.

15. L. E. Schulz and J. Sommerville, "God does not play dice: Causal determinism and preschoolers' causal inferences." *Child Development*, 77(2), 427–442, 2006.
16. http://www.scholastic.com/clifford
17. http://www.pbskidsplay.org/flash2/default.php?page=welcome&lang=en
18. http://www.etc.cmu.edu
19. http://www.fas.org
20. http://www.gamepipe.usc.edu/USC_GamePipe_Laboratory/Home.html
21. http://www.seriousgames.org
22. http://icampus.mit.edu/projects/gamestoteach.shtml
23. http://dmc.utexas.edu
24. http://playgen.com/
25. http://inquiry.uiuc.edu/
26. http://www.mindark.com/
27. http://ipp.gsfc.nasa.gov/mmo/

4

Nobel Laureates Are Role Models in Teaching Nanoscience

> In science the credit goes to the man who convinces the world, not the man to whom the idea first occurs.
>
> **Sir Francis Darwin (1848–1925), *Eugenics Review,* April 1914**

Richard P. Feynman, 1918–1988

Nobel Prize in Physics 1965

Richard P. Feynman (1918–1988) developed a new formulation of quantum theory based, in part, on diagrams he invented to help him visualize the dynamics of atomic particles. In 1965, this noted theoretical physicist, enthusiastic educator, and amateur artist was awarded the Nobel Prize in Physics.

Professor Feynman truly understood the reason for studying science and math, which he tried to explain throughout his lifetime. Feynman was an excellent teacher who enjoyed teaching physics as much as he enjoyed his research. Why and why not? These were always questions he encouraged and expected from all students.

In 1959 his now famous speech, "There is Plenty of Room at the Bottom,"[1] Feynman issued a challenge to physicists to explore the world of atoms and see if they could fit an entire encyclopedia on the head of a pin. He ended the talk with a suggestion to involve high school students in a competition to get them interested in this very small size of scientific inquiry.

> *High School Competition*
>
> Just for the fun of it, and in order to get kids interested in this field, I would propose that someone who has some contact with the high schools think of making some kind of high school competition. After all, we haven't even started in this field, and even the kids can write smaller than has ever been written before. They could have competition in high schools. The Los Angeles high school could send a pin to the Venice high

> school on which it says, "How's this?" They get the pin back, and in the dot of the "i" it says, "Not so hot."
>
> Perhaps this doesn't excite you to do it, and only economics will do so. Then I want to do something; but I can't do it at the present moment, because I haven't prepared the ground. It is my intention to offer a prize of $1,000 to the first guy who can take the information on the page of a book and put it on an area 1/25,000 smaller in linear scale in such manner that it can be read by an electron microscope.
>
> And I want to offer another prize—if I can figure out how to phrase it so that I don't get into a mess of arguments about definitions—of another $1,000 to the first guy who makes an operating electric motor—a rotating electric motor which can be controlled from the outside and, not counting the lead-in wires, is only 1/64 inch cube.
>
> I do not expect that such prizes will have to wait very long for claimants.

Of course that challenge still has not reached the high schools, but it does give us some insight into the mind of a genius who was also an excellent teacher. Amazingly, his motor challenge was quickly met by William McLellan, a meticulous craftsman, using conventional tools; the motor met the conditions, but did not advance the art.[2] In 1985, Tom Newman, a Stanford grad student, successfully reduced the first paragraph of A Tale of Two Cities by 1/25,000, and collected the second Feynman prize. Professor Feynman was a visionary with a brilliant sense of humor and one of the few physicists that could write a science book that would become a national best seller, as did *"Surely You're Joking, Mr. Feynman!": Adventures of a Curious Character,*[3] which was followed by a second book titled *What Do You Care What Other People Think?: Further Adventures of a Curious Character.*[4] The first page of this book is a very enlightening encapsulated view of Professor Feynman's view of the melding of art and scientific inquisitiveness.

> I have a friend who's an artist, and he sometimes takes a view I don't agree with. He'll hold up a flower and say, "Look how beautiful it is," and I'll agree. But then he'll say, "I, as an artist, can see how beautiful the flower is. But you, as a scientist, take it all apart, and it becomes dull." I think he's kind of nutty.
>
> First of all, the beauty that he sees is available to other people—and to me, too, I believe. Although I might not be quite as refined aesthetically as he is, I can appreciate the beauty of a flower. But at the same time, I see much more in the flower than he sees. I can imagine the cells inside, which also have a beauty. There's beauty not just at the dimension of one centimeter; there's also beauty at a smaller dimension.
>
> There are the complicated actions of the cells, and other processes. The fact that the colors in the flower have evolved in order to attract insects to pollinate is interesting; that means insects can see colors. That adds a question: does this aesthetic sense we have also exist in lower forms of life? There are all kinds of interesting questions that come from a knowl-

> edge of science, which only adds to the excitement and mystery and awe of a flower. It only adds. I don't understand how it subtracts.

This opening page says more about the genius of Richard P. Feynman and his ability to excel as a teacher, by always challenging his students to look deeper into nature and ask more questions. A perfect example was the year he spent teaching science in Brazil at the request of the government.

After teaching the first semester, Mr. Feynman realized that the students did not learn any real science, they just memorized facts to pass the test. Therefore, when he was invited by the students to give a review of his experiences of teaching in Brazil, he asked if he could speak candidly, without any limits, and they agreed. This excerpt from *"Surely You're Joking Mr. Feynman!": Adventures of a Curious Character* will give you insight into his personality.

> As the lecture hall was full, he started out by defining science as an understanding of the behavior of nature. Then he asked, "What is a good reason for teaching science?, allowing of course, that no country can consider itself civilized unless...
>
> Then he stated that, "The main purpose of my talk is to demonstrate to you that NO science is being taught in Brazil!"
>
> He went on to point out that he was very excited upon arriving in Brazil, that he noticed so many young elementary school students were buying books on physics, as they do not teach physics to young children in the United States. However, the reason he found that amazing was that you do not find many physicists in Brazil...and he was wondering...Why is that? So many kids are working so hard and nothing comes of it.
>
> Then he held up the elementary physics textbook they were using. "There are no experimental results mentioned anywhere in this book, except in one place where there is a ball, rolling down an inclined plane, in which it says how far the ball got after one second, two seconds, three seconds, and so on. The numbers have 'errors' in them—that is if you look at them, you think you're looking at experimental results, because the numbers are a little above, or a little below, the theoretical values. The book even talks about having to correct the experimental errors—very fine. The trouble is, when you calculate the value of the acceleration constant from these values, you get the right answer. But a ball rolling down an inclined plane, if it is actually done, has an inertia to get it to turn, and will if you do the experiment, produce five-sevenths of the right answer, because of the extra energy needed to go into the rotation of the ball. Therefore, this single example of experimental 'results' is obtained from a fake experiment. Nobody had rolled such a ball, or they would never have gotten those results."
>
> "I have discovered something else," he continued. "By flipping the pages at random, and putting my finger in and reading the sentences on that page, I can show you what's the matter—how it's not science, but memorizing, in every circumstance."

> ...another example...he stuck his finger in and began to read: "'Triboluminescence. Triboluminescence is the light emitted when crystals are crushed...' and there, have you got science? NO!"
>
> "You have only told what a word means in terms of other words. You haven't told anything about nature—what crystals produce light when you crush them, why they produce light. Did you see any student go home and try it? He can't."
>
> "But if, instead, you were to write, 'When you take a lump of sugar and crush it with a pair of pliers in the dark, you can see a bluish flash. Some other crystals do that too. Nobody knows why. The phenomenon is called 'triboluminescence.' Then someone will go home and try it. Then there's an experience of nature."

Reading this explanation by such an honored and respected physicist was heartwarming. He also included a chapter on his experience with the State Board of Education in California, which requested that he serve on the State Curriculum Commission, which had the task of choosing new textbooks for the entire state.

To make a long story short, he ended up with a seventeen-foot bookshelf full of new math textbooks, which he agreed to review for the state. It was a pretty big job, but he read every one of them, exploding like a volcano every so often because he felt the books were so lousy. As he stated, "They were false, they were done hurriedly," and he felt everything was a little bit ambiguous—"they weren't smart enough to understand what was meant by 'rigor.'"

The books were so bad that the commission ended up recommending supplementary books as a package to help the teachers. In the end the whole project was scrapped as the board of education did not have enough money passed by the senate to purchase the recommended books. The following year they were going to review science textbooks, and Mr. Feynman did look at a few of them, but they all turned out to be equally horrifying, which cinched his decision to resign from the commission. The saddest part of this story is the fact that these events took place in the decade of the 1960s, and nothing has really changed. I think if Professor Feynman were alive, he would ask: "Why have they still not addressed the problems I pointed out in the '60s?"

The last chapter of the book is adapted from the 1974 Caltech Commencement Address in which Mr. Feynman addressed "integrity in science and in taking our place in the world." The closing remarks tell us so much about the world view of Richard P. Feynman ... the man who enjoyed the simple pleasure of finding things out. No ordinary genius, but he was an exemplary role model in these troubled times as we struggle with the lack of good education in our schools.

> So I have just one wish for you—the good luck to be somewhere where you are free to maintain the kind of integrity I have described, and

> where you do not feel forced by a need to maintain your position in the organization, or financial support, or so on, to lose your integrity. May you have that freedom.
>
> **Richard P. Feynman 1974**

Perfect Reasonable Deviations from the Beaten Track: The Letters of Richard P. Feynman,[5] edited by Michelle Feynman, is the most recent book about Feynman to be published. The review written upon release says it all so eloquently:

> En route to a conference on liquefied helium and high-energy physics, Richard Feynman wrote to his young niece describing the work that scientists do. "Atoms are complicated," he explained in a letter datelined "flying over England." "Maybe like watches are—but atoms are so small that all we can do is smash them together and see all the funny pieces (gears, wheels, and springs) which fly out. Then we have to guess how the watch is put together… Now it looks like we know most of the parts that go in—but nobody knows how they fit together."
>
> Feynman won the Nobel Prize in Physics in part for figuring out how all those parts that go in fit together. Technically, in the words of the Swedish Royal Academy, he won it for "fundamental work in quantum electrodynamics with deep-ploughing consequences for the physics of elementary particles."
>
> Feynman was already on his way to minor celebrity before the prize. His Lectures on Physics had brought him great acclaim but television made him famous. "Dear Richard," wrote one swooning fan, "I've fallen in love with you from seeing you on NOVA." Only Captain Kirk could make time travel sound sexier. But Kirk could only say, "Beam me up." Feynman could actually explain it.
>
> Feynman could be testy, particularly when someone wrote to him with a question without thinking hard about it first. But he was also short with anyone who questioned the value of scientific inquiry. After Feynman had disparaged modern poets for a lack of curiosity, an admirer sent him a copy of Auden's "After Reading a Child's Guide to Modern Physics" and invited him to recant. "Mr. Auden's poem," Feynman wrote in response, "only confirms his lack of response to Nature's wonders for he himself says that he would like to know more clearly what we 'want the knowledge for.' We want it so we can love Nature more. Would you not turn a beautiful flower around in your hand to see it from other directions as well?" By putting science in the service of beauty and awe, the ever-romantic Feynman beats the poets at their own game. Wonder and imagination were his main tools. Particle-accelerators and electron-microscopes just made the job easier.[5]

The *Vegas Science Trust*[6] in the United Kingdom has posted a series of four videos from Auckland University, New Zealand, that teachers can stream to their classrooms. The videos were chosen by *The New Scientist*[7] as the best online videos in 2007. They were filmed from the Douglas Robb Memorial Lectures on *Photons—Corpuscles of Light, Fits of Reflection and Transmission—Quantum Behavior, Electrons and Their Interactions, New Queries, What Does It Mean and Where Is It All Leading?*

Richard Errett Smalley (1943–2005)

Nobel Prize in Chemistry 1996

Awarded to Richard E. Smalley and Robert F. Curl, both supported by the Office of Science, Rice University and Curl's colleague Sir Harold W. Kroto of Great Britain, for their discovery of C_{60}—a new class of carbon structures (see Figure 4.1).

Scientific Discoveries Follow Multiple Paths of Inquiry

Several lines of research—in spectroscopy, astronomy, and metallic clusters—converged in 1985 to lead to the discovery of an unusual molecule. This cluster of 60 carbon atoms was especially stable because of its hollow, icosahedral structure in which the bonds between the atoms resembled the patterns on a soccer ball. The molecule was named buckminsterfullerene after the geodesic domes designed by architect Buckminster Fuller. The identification of this form of carbon (also called buckyballs) sparked broad interest in the chemistry of an entire class of hollow carbon structures, referred to collectively as fullerenes. Formed when vaporized carbon condenses in an atmosphere of inert gas, fullerenes include a wide range of shapes and sizes, including nanotubes of interest in electronics and hydrogen storage. The initial discovery was recognized by the 1996 Nobel Prize in Chemistry, awarded to Richard E. Smalley and Robert F. Curl, both supported by the

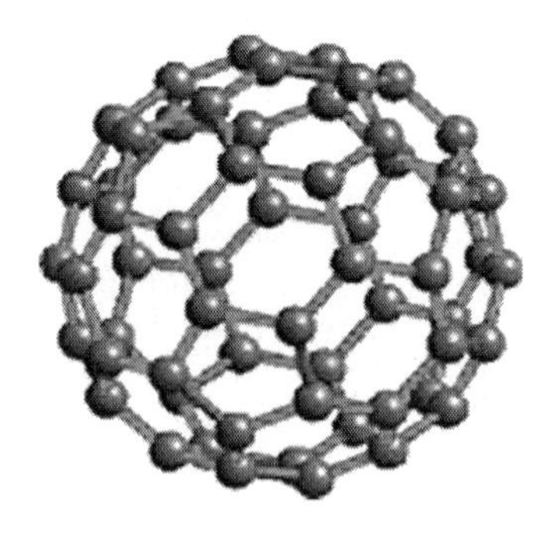

FIGURE 4.1
A new allotrope of carbon that consists of 60 carbon atoms in the shape of a soccer ball.

Office of Science, and Curl's colleague Sir Harold W. Kroto of Great Britain. More recently, scientists at Lawrence Berkeley National Laboratory reported a new synthetic method for producing, extracting, and purifying a cluster of 36 carbon atoms in quantities useful for research purposes; they also confirmed the high reactivity and other unusual electrical and chemical properties of this material.

The scientific impact of the discovery of fullerenes launched a new branch of chemistry, and related studies have contributed to growing interest in nanostructures in general and the principles of self-assembly. Fullerenes also have influenced the conception of diverse scientific problems such as the galactic carbon cycle and classical aromaticity, a keystone of theoretical chemistry.

The social impact of the discovery of fullerenes has not fully been realized because they are highly versatile (there are literally thousands of variations) and thus have many potential applications. For example, fullerene structures can be manipulated to produce superconducting salts, new three-dimensional polymers, new catalysts, and biologically active compounds.[8]

Long ago, the eighteenth century mathematician Leonhard Euler established that every closed polygon made with hexagons and pentagons must contain exactly 12 pentagons. C_{60}'s soccer-ball shape is the smallest possible structure in which the 12 do not touch; in any smaller structure, the pentagons must touch.[9]

Naming the Buckminsterfullerene[10]

When Harry Kroto and Richard Smalley, the experimental chemists who discovered C_{60} named it buckminsterfullerene, they accorded to Richard Buckminster Fuller (1895–1985), the American engineering and architectural genius, a special type of immortality that only a name can confer—particularly when it links a single historical person to a hitherto unrecognized universal design in the material world of nature: the symmetrical molecule C_{60}. Smalley's laboratory equipment could only tell them how many atoms there were in the molecule, not how they were arranged or bonded together. From Fuller's model they intuited that the atoms were arrayed in the shape of a truncated icosahedron—a geodesic dome. Only after novel phenomenon or concept is named can it be translated into the common currency of thought and speech. [See Figure 4.2.]

This newly discovered molecule, a third allotrope of carbon—ancient and ubiquitous—transcends the historical or geographical significance of most named phenomena such as mountains of the moon or Antarctic peaks and ridges. Cartographers named two continents for Amerigo Vespucci, because he asserted (as Columbus did not) that the coasts of Brazil and the islands of the Caribbean were a landmass of their own and not just obstacles on the route to Asia. C_{60} is a far more elemental discovery; it is more ancient; and it pervades interstellar space. Fuller has no reason to envy Vespucci.

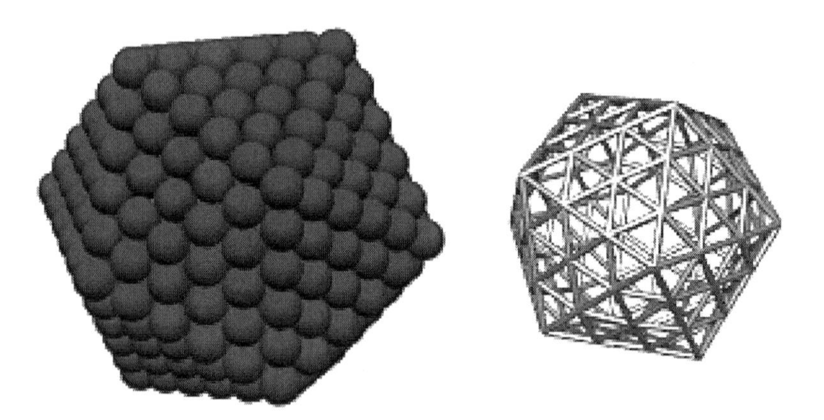

FIGURE 4.2
Truncated icosahedron.

Buckminsterfullerene was discovered by chemists who were not looking for what they found. Kroto was looking for an interstellar molecule. Smalley said he hadn't been very interested in soot, but they agreed to collaborate. Smalley's laboratory at Rice University had the exquisite laser-vaporization and mass spectrometry equipment to describe the atoms of newly created molecules. Scientific experimenters investigate nature at a level where revelation is often unpredictable and sometimes capricious. This is a phenomenon that Fuller (who was not a scientist, but a staunch defender of the scientific method) generalized into the dogmatic statement that all true discovery is precessional. For Fuller, the escape from accepted paradigms is precessional. (Vespucci precessed; Columbus did not.) Fuller had a lifelong preoccupation with the counter-intuitive, gyroscopic phenomenon of precession. He defined precession, quite broadly, as the effect of bodies in motion on other bodies in motion. Every time you take a step, he said many times, you precess the universe.

For that matter, one may say that Kroto and Smalley, in recognizing the shape of the C_{60} molecule made a precessional discovery. Earlier, Osawa, in a paper published in Japanese in 1970, had described the C_{60} molecule with the truncated icosahedral shape; so had Bochvar and Gal'pern in 1975 when they published a paper in Russian on the basis of their calculations. They all recognized the novelty of the molecule and conjectured that its structure should afford great stability and strength. However, neither Osawa nor Bochvar and Gal'pern had experimental evidence, nor did they consider their result important enough to follow up their finding with further work or to convince others to do so. Curiously, in 1984 a group of Exxon researchers made an experimental observation of C_{60} along with many other species. They failed, however, to discern the shape of this species and did not recognize its special importance. These precursors to Kroto and Smalley apparently lacked the requisite—precessional—insight to appreciate the significance of what they had found. Kroto and Smalley's precessional insight was best manifested by their decision to give a name to the C_{60} molecule of the truncated icosahedral shape.

After a few letters objecting to the name buckminsterfullerene had appeared in the columns of Nature,[11] Harry Kroto gallantly defended its choice on the grounds that no other name—none of the forms of the classic Greek geometers—described the essential three properties of lightness, strength, and the internal cavity that the geodesic dome affords. To the protest that nobody had ever heard of Fuller, he submitted that the name would have educational value. A fine exercise of onomastic prerogative.

Fuller was not a chemist. He was not even a scientist, and made no pretension of adhering rigidly to an experimental and deductive methodology, and he did not follow the rules of submitting published papers to peer review. But he had an extraordinary facility for intuitive conceptioning. Jim Baggott, in his superb account *Perfect Symmetry: The Accidental Discovery of Buckminsterfullerene*[12] quotes Fuller in an epigraph: "Are there in nature behaviors of whole systems unpredicted by the parts? This is exactly what the chemist has discovered to be true." Baggott goes on to describe how Fuller had derived his vector equilibrium (cuboctahedron, in conventional geometry) from the closest packing of spheres of energy. What he had was a principle that led to the design of geodesic structures capable of a strength-to-weight ratio impossible in more conventional structures. Fuller had a highly generalized definition of the function of architecture that put him outside the scope of the academicians' view of their discipline. Bucky said "architecture is the making of macrostructures out of microstructures."

Baggott concludes: "Fuller's thoughts about the patterns of forces in structures formed from energy spheres had led him to the geodesic domes.... That his geodesic domes should serve as a basis for rediscovering these principles in the context of a new form of carbon microstructure has a certain symmetry that Fuller would have found pleasing, if not very surprising."[13]

How Discoveries Transition

The discovery of C_{60} buckminsterfullerenes, and their next transitional phase into carbon nanotubes, was stimulated by many other researchers and friends. Since winning the Nobel Prize in 1996, the late Dr. Smalley's work garnered 3,816 total citations for 78 papers, making him the most-cited scientist of nanotechnology research in the past decade. Dr. Smalley was the Gene and Norman Hackerman Professor of Chemistry and Professor of Physics at Rice, as well as the Director of the Center for Nanoscale Science & Technology at Rice.

Transition from Buckminsterfullerene to Carbon Nanotubes

Dr. Smalley continued to work with buckminsterfullerene and, by 1990, firmly realized that if you had n carbons in a fullerene—n being a number bigger than 60—the most stable form of that structure would be as ball-shaped as possible. He also knew there was a coalescence of balls to make larger balls and that coalescence almost certainly would be more active at the ends. Since this is where the pentagons were, it tends to make it a more elongated object.

In December 1990, Wolfgang Krätschmer and Donald Huffman had their major breakthrough making fullerenes in a carbon smoke generator. While attending a workshop organized by the Air Force Office of Scientific Research on the subject of carbon–carbon materials, Dr. Smalley was asked to be on the panel. He decided to discuss the fact that if you make an elongated fullerene—the way C_{70} is the same as C_{60}, but with a belt of carbons around the equator to make it elongated—by adding more and more belts, the result would be a long tube with hemispheres of C_{60} as end caps. It would be a fullerene carbon fiber with the virtue of having no exposed edges. It is the exposed edges of carbon fibers that is their Achilles heel, where oxidation primarily occurs and also where fractures occur, which was the topic of discussion during the meeting.

Millie Dresselhaus [of MIT] was sitting right next to him during the panel discussion, and over the subsequent months, Millie and Gene Dresselhaus got intrigued with the structure. Soon they started calculating the infrared spectra and the electronic properties. The next summer, Millie gave a talk about bucky tubes and presented the results of their calculations. When he asked her if she had any idea how you could make these things, she responded that he was already making them. That was the origin of carbon nanotubes.

The Nobel Laureates are inspirational role models for young students, who might benefit from learning the personal stories of these scientists and what prompted their early interest in scientific studies. The Nobel Prize[13] Web site has the autobiographies of many Nobel winners, along with games and other educational components that teachers can use in the classrooms.

The autobiographies[14] section on Richard Smalley, as an example, provides some insight on his first exposure to science as a young child.

> It was from my mother that I first learned of Archimedes, Leonardo da Vinci, Galileo, Kepler, Newton, and Darwin. We spent hours together collecting single-celled organisms from a local pond and watching them with a microscope she had received as a gift from my father. Mostly we talked and read together. From her I learned the wonder of ideas and the beauty of Nature (and music, painting, sculpture, and architecture).

Nobelprize.org has a unique way of introducing the Nobel Prize that goes beyond the mere presentation of facts. These introductions, aptly called educational, are made in the form of games, experiments, and simulated environments ready to be explored and discovered. The productions are aimed at the young, particularly the 14 to 18 age group, who may know about the Nobel Prizes and the Nobel Laureates, but often lack a deeper understanding about the Nobel Prize–awarded works.

These educational productions do not require previous knowledge. A central thought or issue is explored during 10 to 20 minutes of activity, using a specific Nobel Prize-awarded work as a springboard for the whole exercise.

The productions offer an excellent way of using the Internet for homework, or just plain, wholesome entertainment. The high level of interactivity and the sophisticated illustrations ensure an enriching time spent in front of the computer.

Leon M. Lederman, 1922

Nobel Prize Physics 1988

Leon Lederman, Batavia, Illinois, USA, Melvin Schwartz, Mountain View, California, USA, and Jack Steinberger, Geneva, Switzerland, jointly received the Nobel Prize for Physics for the neutrino beam method and the demonstration of the doublet structure of the leptons through the discovery of the muon neutrino.

The experiment was planned when the three researchers were associated with Columbia University in New York, and carried out using the Alternating Gradient Synchrotron (AGS) at Brookhaven National Accelerator Laboratory on Long Island, New York. Leon Lederman retired as Director of the Fermi National Laboratory in Batavia, near Chicago, Illinois, where the world's largest proton accelerator is situated. Melvin Schwartz, formerly professor at Columbia and Stanford Universities, is now president of his own firm specializing in computer communications, in Mountain View, California. Jack Steinberger, who is an American citizen, has worked as a Senior Physicist at CERN, Geneva, Switzerland, where he has led a number of large experiments in elementary particle physics, including experiments that employ neutrino beams.

The work rewarded by the Nobel Committee was carried out in the 1960s. It led to discoveries that opened entirely new opportunities for research into the innermost structure and dynamics of matter. Two great obstacles to further progress in research into weak forces—one of nature's four basic forces—were removed by the prizewinning work. One of the obstacles was that there was previously no method for the experimental study of weak forces at high energies. The other was theoretically more fundamental and was overcome by the three researchers' discovery that there are at least two kinds of neutrino. One belongs with the electron, the other with the muon. The muon is a relatively heavy, charged elementary particle which was discovered in cosmic radiation during the 1930s. The view, now accepted, of the paired grouping of elementary particles has its roots in the prizewinners' discovery.

What Are Neutrinos?

Neutrinos are almost ghostlike constituents of matter. They can pass unaffected through any wall; in fact all matter is transparent to them. During the conversion of atomic nuclei at the center of the sun, enormous quantities of neutrinos (which belong to the electron family) are produced. They pass through the whole sun virtually unhindered and stream continually from its surface in all directions. Every human being is penetrated by sun neutrinos at a rate of several billion per square centimeter per second, day and night, without their leaving any noticeable trace. Neutrinos are inoffensive. They have no electrical charge, and they travel at the speed of light, or nearly. Whether they are weightless or have a finite but small mass is one of today's unsolved problems.

The contribution awarded consisted, among other things, of transforming the ghostly neutrino into an active tool of research. As well as in cosmic radiation, neutrinos that belong to the muon family can be produced in a multistep process in particle accelerators, and this is what the prizewinners utilized. Suitable accelerators exist in a few laboratories throughout the world. Since all matter is transparent to neutrinos, it is difficult to measure their action. Neutrinos are, however, not wholly inactive. In very rare cases a neutrino can score a random direct hit or, more correctly, a near miss, on a quark, a pointlike particle within a nucleon (proton or neutron) in the nucleus of an atom or on a similarly infinitesimal electron in the outer shell of an atom. The rarity of such direct hits implies that a single neutrino of moderate energy would be able to pass unhindered through a wall of lead of a thickness measured in light-years. In neutrino experiments the rarity of the reactions is compensated for by the intensity of the neutrino beam. Even in the first experiment, the number of neutrinos was counted in hundreds of billions. The probability of a hit also increases with the energy of the neutrinos. The method of the prizewinners makes it possible to achieve very high energies, limited only by the performance of the proton accelerator. Neutrino beams can reveal the hard inner parts of a proton in a way not dissimilar to that in which X-rays reveal a person's skeleton.

When the neutrino beam method was invented by the Columbia team at the beginning of the 1960s the quark concept was still unknown, and the method has only later become important in quark research. Also of later date is the experimental discovery of an entirely new way for a neutrino to interact with an electron or a quark, in which it retains its own identity after impact. The classical way of reacting implied that the neutrino was converted into an electrically charged lepton (electron or muon), and this was the reaction utilized by the prizewinners.

The Prizewinners' Experiment

The very first experiment using a beam of high-energy neutrinos originated in one of the daily coffee breaks at the Pupin laboratory, where faculty and research students would relax together for half an hour. In this stimulating atmosphere around Nobel Prizewinners T. D. Lee, C. N. Yang (Nobel Prize for Physics 1957), and others at the end of the 1950s, the need to find a feasible method of studying the effect of weak forces at high energies was discussed. Hitherto it had only been possible to study processes of radioactive decay, spontaneous processes at necessarily relatively low energies. Beams of all common particles (electrons, protons, and neutrons) were discussed. While these are relatively simple to produce, they were found to be unusable for this purpose. The apparently hopeless situation suddenly changed when Melvin Schwartz proposed that it ought to be possible to produce and use a beam of neutrinos. During the next two years he, together with Leon Lederman and Jack Steinberger, worked on the proposal in order to achieve a sufficiently intense beam of neutrinos free from all other types of particle, and to design a detector for measuring neutrino reactions. The group at Columbia also included G. Danby, J. M. Gaillard, K. Goulianos, and N. Mistry.

The neutrinos in the Columbia experiment were produced in the decay in the flight of charged pi-mesons. In a first step, protons were accelerated to high velocities and directed at a target of the metal beryllium. As the next step, high-velocity pi-mesons were produced in a forward-directed beam. Mesons are radioactive, and they decayed into a muon and a neutrino each when allowed to travel a path of free flight, which was set at 21 meters. In this step high-energy neutrinos were produced as a forward-directed beam, still containing quantities of leftover pi-mesons and muons which had been formed at the same time. To eliminate these unwanted particles completely from the beam, a 13.5-meter-thick wall of steel was needed. The material came from scrapped warships. The measuring device (detector) was built behind the wall, which of course was transparent to the neutrinos. So that the detector should not be entirely transparent, it was thought best to build it as a 10-ton spark chamber, then a new and fairly untested type. The detector consisted of a large number of aluminum plates with spark gaps between them. A muon or an electron produced by a neutrino in one of the aluminum plates photographed its own track as a series of sparks, using a special self-exposing device.

A burning problem had arisen at the time of the experiment regarding the measurements of muon radioactive decay. The measurement results, to which Jack Steinberger and Bruno Pontecorvo, among others, contributed, disagreed with accepted theoretical calculations. The problem was addressed by many researchers, among them G. Feinberg and T. D. Lee, as well as methodologically by Pontecorvo, and they indicated that one way out of the dilemma would be the existence of two entirely different types of neutrino.

If the neutrinos in the Columbia experiment beam were identical to the neutrinos common in beta decay, the reactions in the detector should convert the neutrino to a fast electron as often as to a fast muon. On the other hand only muons would result if there were two different kinds of neutrino. The prizewinners and their collaborators arranged their detector so that the cause of the spark tracks could be interpreted. The results showed that only muons were produced by the neutrinos in the beam, no electrons. Thus there exists a new type of neutrino that forms an intimate pair with the muon. Consequently the electron forms its own delimited family with its neutrino.

The discovery thus had immediate consequences. Knowledge of the role of the family concept and the great importance of the method within elementary particle physics has grown during the time that has elapsed since the discovery was made. A question that is still current is whether or not small departures from strict family membership occur.

Leon M. Lederman retired from Fermilab in 1989 to join the faculty of the University of Chicago as professor of physics. In 1989, he was appointed Science Adviser to the Governor of Illinois and helped to organize a Teachers' Academy for Mathematics and Science, designed to retrain 20,000 teachers in the Chicago Public Schools in the art of teaching science and mathematics. In 1991 he became president of the American Association for the Advancement of Science.

The following story is my personal journey to meet Dr. Lederman, who is such an inspirational speaker and teacher. I observed him interacting with the parents and students in the audience, challenging them to continue inquiring about everything in nature. As the audience became reverently silent, he added some humor by describing his own high school record as a mediocre C-student, stating that his curiosity in scientific inquiry came later—reinforcing hope for students who seem to lack interest.

An Evening with Leon Lederman, Nobel Laureate, at The Bakken Museum

Accepting an invitation to spend the day in Minneapolis, Minnesota, on December 8, 2005, was a wise decision. Having just participated in the *Beyond Einstein*[15] webcast from CERN, I was very interested in meeting Leon Lederman in person. His participation from the Fermilab in Illinois was a humorous, educational parody of the David Letterman show titled *Late Night with Lederman,* which ran for 1 hr. 37 min.

The afternoon talk for students by Earl Bakken, the founder of The Bakken Museum, and Leon Lederman was aimed at stimulating the young students in the audience to seek more knowledge in science and math. Their premise—science basically teaches a young mind "how the world works"—was that it should stimulate their natural curiosity to investigate the world around them.

Some of the parents in the audience were inquiring about mentors for their young children to keep their passion for scientific curiosity expanding. They said schools were not teaching science to elementary students in grades K–3, and their children were attempting to learn on their own. One child, who came to all the events at the museum, had built his own science project at home. At the end of the program he presented it to Earl Bakken as a gift. He was only seven years old. The afternoon session set the tone for the evening lecture that was to be presented by Leon Lederman, titled *The Quiet Crisis*.

The Quiet Crisis

Lederman had been working with educational leaders for many years and was passionate about the programs he was initiating around the country. It all starts with a question that stimulates parents, teachers, and administrators to take action, and work with the scientists:

Why Must Scientists Be Involved in Education and What Can Scientific Spirit Contribute?

Dr. Lederman's answer ... they instill the qualities that make science.

What is Scientific Thinking?

- Blend of curiosity and ego
- Humility in relationship to the heritage
- Skepticism about universal validity of what we learned
- A liberating sense of freedom to question authority
- An open mindedness regarding new ideas
- A confidence in rationalization
- Faith in the beauty of science

Saturday Morning Physics (a short-course for high school students) was initiated by Lederman in 1980. He has been an outspoken advocate for new approaches to secondary science that emphasize a coherent three-year science curriculum beginning with physics. There are a growing number of schools introducing the new curricula inspired by his advocacy.

Leon Lederman feels very strongly that the academic and corporate worlds need to respond to the growing concern that the United States is now trailing other nations in producing scientists, engineers, and mathematicians. As I listened to this brilliant man, I was so very glad I accepted the invitation.

Lederman described science as a way of thinking and that we needed to make sure every student graduating from high school has it. He felt it was time to wage war on ignorance stating, "We cannot maintain a twenty-first century economy with a nineteenth century curriculum." He alluded to the National Academies report, *Rising Above the Storm,* which states that we need to train 10,000 teachers every year at a cost of $20,000 per teacher.

There are 15,000 school districts in the United States without a coherent strategy. Therefore over the past 10 years he has developed a program titled *America's Renaissance in Science—Physics First*[16] (ARISE), which proposes a core curriculum for a coherent three-year science sequence, titled simply: *Science I, Science II, Science III.* Lederman stated that we teach backwards. Mathematics is the base foundation of all science, and we do not introduce algebra into our current curriculum until 9th grade. We also introduce biology in 9th grade, chemistry in 10th grade, and physics is introduced last in 12th grade. This decision was made in 1893, so we have been teaching backwards for over a hundred years.

There is a hierarchy in science that depends on mathematics as the base for all other subjects. Algebra I should be introduced in 7th grade, along with conceptual physics, which avoids the use of extensive math, but emphasizes a grasp of concepts.

Example:

$$X = VT \text{ (velocity} \times \text{time)}$$

The concepts of momentum, force, acceleration, temperature, etc., along with the conservation laws are the natural base of *Physics First*, which can be told with storytelling. *Physics First* teaches that the key to modern science is the atom: "everything is made of atoms." We need to be teaching the structure of atoms in 7th grade, vertical to the concepts of the Introduction to Algebra. This makes room for advanced courses in physics at the high school level. Chemistry is next in 8th grade, because the teachings in chemistry depend on quantum theory at the atomic level, which are covered in physics. Biology is third in the pyramid, because cells are made from molecules, which are composed of atoms.

After the introductory courses in middle schools, the same hierarchy can be followed for a deeper comprehension of the basic sciences in high school, and the inclusion of higher levels of mathematics, such as geometry, trigonometry, and Calculus I and II. By teaching the basic introduction to Algebra, Algebra I and Algebra II in middle school, our students can be competitive globally and prepared for their college courses. Our high school students currently need one math course and one science course per year in high school to graduate. Globally the requirements are three science and three math courses per year to graduate high school. Even if we start to make these changes now, it will be a couple of decades before our graduates are able to compete and excel in science and math.

As of 2005, Lederman had 1000 high schools out of over 24,000 nationwide teaching *Physics First.* Many of them are in San Diego, California. Lederman enthusiastically ended the talk with the statement, "It Works!"

Sciences Need to Be Taught As a Humanistic Activity

How does it work? How messy is the process of discovery? We need to teach by experimenting with new ways of communicating and integrating our subjects with a coherence that makes sense in the twenty-first century. Team teaching and learning through collaborative efforts encourage students to apply the concepts they are learning about their world to actual situations in their lives. The world is not flat, and the crisis is real!

References*

1. http://www.zyvex.com/nanotech/feynman.html
2. Richard Phillips Feynman and Christopher Sykes, *No Ordinary Genius: The Illustrated Richard Feynman*, p. 175, W. W. Norton & Company, 1995, ISBN 039331393X, 9780393313932.
3. Richard P. Feynman, *"Surely You're Joking, Mr. Feynman!": Adventures of a Curious Character*, Paperback, 1997, ISBN 0-393-31604-1.
4. Richard P. Feynman, Ralph Leighton, *What Do You Care What Other People Think?: Further Adventures of a Curious Character*, 2001, ISBN 10: 0-393-32092-8.
5. Michelle Feynman (Ed.), *Perfectly Reasonable Deviations from the Beaten Track: The Letters of Richard P. Feynman*, Basic Books, ISBN: 0738206369, 384 pages.
6. http://www.vega.org.uk/video/subseries/8
7. http://www.newscientist.com
8. H. W. Kroto, J. R. Heath, S. C. O'Brien, R. F. Curl, and R. E. Smalley, C_{60}: Buckminsterfullerene, *Nature* 318, 162, November 14, 1985.
9. http://www.lbl.gov/Science-Articles/Archive/carbon-36-superconductor.html.
10. http://bfi.org
11. H. W. Kroto, *Nature (London)*, 329, 529, 1987.
12. J. Baggott, *Perfect Symmetry: The Accidental Discovery of Buckminsterfullerene*, Oxford University Press: Oxford, 1994.
13. http://nobelprize.org/educational_games/
14. http://nobelprize.org/nobel_prizes/chemistry/laureates/1996/smalley-auto-bio.html
15. http://beyond-einstein.web.cern.ch/beyond-einstein/
16. http://ed.fnal.gov/arise

* All links active as of August 2010.

Section II

Teaching Nanotechnology

> The outcome of any serious research can only be to make two questions grow where only one grew before.
>
> **Thorstein Veblen (1857–1929)**

This section defines nanotechnological literacy and turns to the critical process of teaching K–12 students the skills to understand and evaluate all technologies they may encounter. It is authored by Miguel F. Aznar, Educational Director at the Foresight Institute, who teaches high school students nanotechnology in COSMOS, the California State Summer School for Mathematics and Science, at the University of California at Santa Cruz.

5

What is Nanotechnological Literacy?

> There is a major difference between technological competence and technological literacy. Literacy is what everyone needs. Competence is what a few people need in order to do a job or make a living. And we need both.
>
> **William Wulf**

Imagine a robotic submarine small enough to slip through the smallest capillary (<4 micrometers) to patrol our bloodstream. Programmed to identify specific bacteria or viruses, it would grab them with tiny telescoping arms and stuff them into its mouth, chop them into pieces, and digest them before ejecting the harmless components. When the nanoscale robots, or "nanobots," have done their job, they can be filtered out of the blood. Robert Freitas has imagined just such medical devices, naming them "microbivores." In a research paper,[1] Freitas draws on many disciplines to answer questions about microbivores (Figure 5.1):

- **Immunology**: What would not provoke our immune system?
- **Physiology**: What shape and propulsion system would work in our circulatory system?
- **Materials science**: How could microbivores' surfaces be compatible with our bodies' immune and circulatory systems?
- **Probability:** How long will it take for microbivores to bump into pathogens?
- **Cell biology and bacteriology**: How do we positively identify specific pathogens?
- **Physics**: How much force is required to drag a pathogen through blood plasma to the microbivore's mouth?
- **Mechanical engineering**: What structures could manipulate the pathogens once identified?
- **Electrical engineering**: What energy would be required to operate it, and how would doctors communicate with it?
- **Computer science**: What information processor will fit into the space, energy, and heat limitations of a microbivore?
- **Biochemistry**: What enzymes could break down the pathogens after they have been minced?

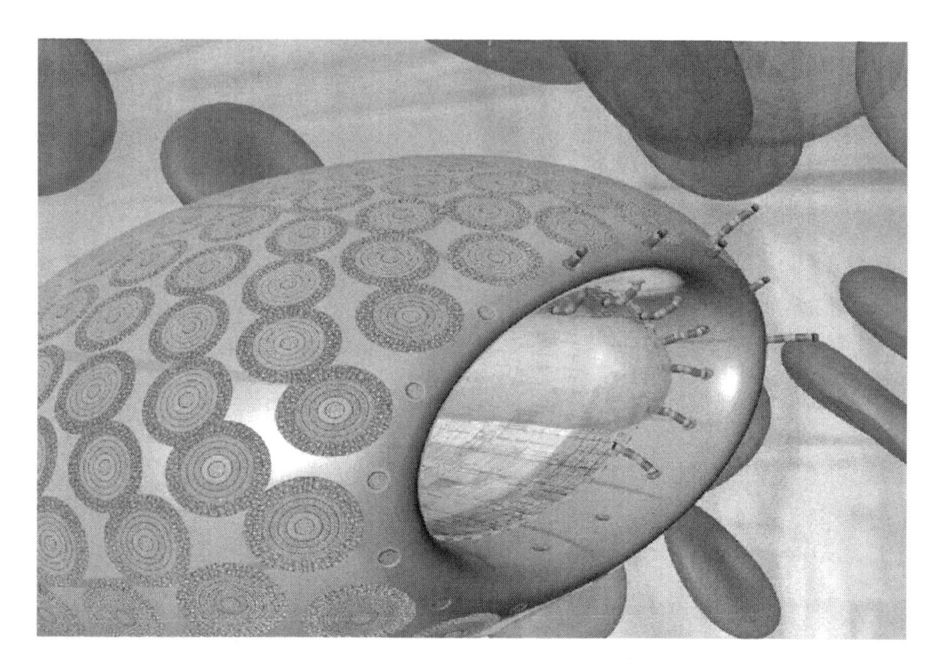

FIGURE 5.1
The microbivore, an artificial white cell, floats along in the bloodstream until it encounters a pathogen. © 2001 Zyvex Corp. and Robert A. Freitas, Jr. Designer Robert Freitas, additional design Forrest Bishop.

- **Medicine**: How do we extract the microbivores from the bloodstream?
- **Risk analysis:** What failure modes are most likely, and which systems need redundancy?

The microbivore—whether or not we ever make it—illustrates the challenge of nanotechnology education. It could be breathtakingly panoramic, and of course, this list is far from complete. We omitted quantum chemistry and nanoethics, among many other fields, any one of which could occupy a bright student for a lifetime. Visionaries may try to span all these fields, but scientists, engineers, and technicians understand a few fields and specialize in one or two. The general public benefits from understanding nanotechnology, just as nanotechnology benefits from their understanding it. Anthropologist Chris Toumey looked to patterns in public reactions to new technologies when he observed:

> That nanotechnology is a blessing or a curse; that scientists can be trusted or should be feared; that all will enjoy its benefits, or that a few will control its powers: these kinds of pre-existing feelings about science will be at least as influential as the scientific merits of the research in shaping public reactions to nanotechnology. The same was true in

> the earlier cases of fluoridation, cold fusion, creationism-versus-evolution, embryonic stem cell research, and many more forms of science and technology.[2]

Do you buy a sunscreen with nanoparticles? Do you choose a cancer treatment that shuttles drugs in nanoscale shells through your bloodstream to the tumor? Do you vote for a politician who promises either public funding for or tighter regulations on nano research? While scientists, engineers, and technicians need specialized knowledge, the public needs a foundation for evaluating new technologies. In the absence of an ability to evaluate nanotechnology, the public will be prone to blind acceptance or rejection. This is not just a domestic problem, but a global one. Nanotechnology's success in the world may depend on widespread nanotechnology education. Former Secretary General of the United Nations Kofi Annan made this point:

> If every nation gains full access to this broader world community of science and has the opportunity to develop an independent science capability, its public can engage in a candid dialogue about the benefits and risks of new technologies, such as genetically engineered organisms or nanotechnology, so that informed decisions can be made about their introduction into our lives.[3]

What a vast gulf exists between creating the nanotechnologies of the future and the knowledge of students in K–12 —or even the knowledge of those who would teach them! While we may inspire, or at least foster, the visionary in K–12 , we will not create her. Nor will we graduate scientists, engineers, or technicians from K–12 . On the other hand, universities will launch only the exceptional student into these careers without the important foundation that K–12 can produce.

In K–12 , what can we teach about nanotechnology? What should we? How about starting with literacy before competency? In the area of nanotechnology, literacy is a grasp of the grand patterns that transcend the details, enabling us to evaluate costs and benefits. And competency is a grasp of those details in one or several specialties, expertise enabling us to imagine, experiment, design, and operate in the nanoscale. Competency may start in K–12 , but the science, engineering, and math that scientists, engineers, and technicians will need requires college (with exceptions for the geniuses).

Many students may have little interest in nanotechnological competency. They will pursue careers other than visionary, scientist, engineer, or technician. Still, they will be confronted with choices about nanotechnology. Literacy about nanotechnology will enable them to think critically about their alternatives. It will prepare them to assess the reports and counsel of experts. Since we all rely on experts in fields beyond our own, and nanotechnology is too vast for anyone to master fully, everyone would benefit from literacy. Experts—visionaries, scientists, engineers, and technicians—have often been accused of an inability to communicate. When explaining to the public why something is not simply possible, but also desirable, they will greatly benefit from literacy because its essence is context. To explain nanotechnology—or anything—we need context. To critically evaluate nanotechnology—or anything—we need context.

Enough talking around literacy and how wonderful it is, what might it look like? One teaching approach that has worked for literacy in both nanotechnology and technology in general is framed by nine simple questions. Used with students from 10 to 18 years old, the approach is called ICE-9 because the nine questions are grouped into Identity, Change, and Evaluation:

Identity:

1. What is technology?
2. Why do we use it?
3. Where does it come from?
4. How does it work?

Change:

5. How does it change?
6. How does it change us?
7. How do we change it?

Evaluation:

8. What are its costs and benefits?
9. How do we evaluate it?

ICE-9 can be presented graphically as a pyramid, with Identity as the foundation on which we build an understanding of Change. That combined foundation allows us to explore Evaluation (Figure 5.2).

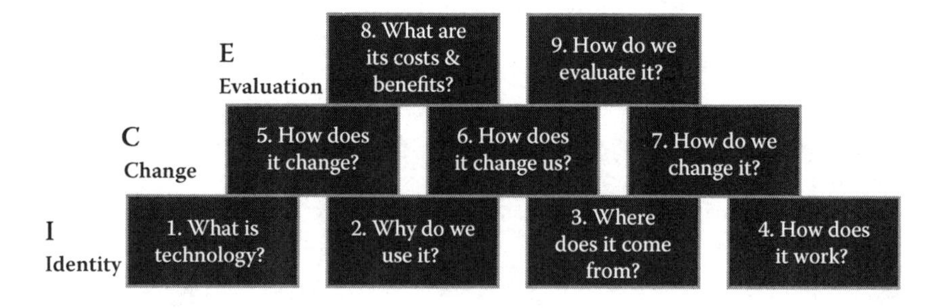

FIGURE 5.2
ICE-9 is an approach for critical thinking about technology. (Source: www.KnowledgeContext.org.)

The questions are clearly broad, ranging from setting the scope of our exploration all the way to reflecting on personal values. If we are to educate today's students for a future we cannot predict, both the questions and the answers we offer must be broad and inclusive. Microbivores are in one possible future, and we want to prepare young people for many futures. As with technology as a whole, nanotechnology follows some simple patterns, making it possible to find answers that will age gracefully into any of a variety of futures.

A further benefit of focusing on literacy before competency is for teachers. Teachers need no technical university degree to illuminate why we use nanotechnology, where it comes from, or the basics of how it works. They need just knowledge of—not expertise in—chemistry, physics, and math to teach how nanotechnology changes or what tradeoffs we tend to make between costs and benefits. Nanotechnology is challenging, in part, because it connects to so many areas. With literacy, we turn this problem into opportunity, inviting teachers to draw on their diverse backgrounds to share the context of nanotechnology for their students.

We learn not simply through broad questions and enduring answers, though those two components are key in the ICE-9 approach. Many learn from stories, so the lesson for each of the ICE-9 questions begins with stories illustrating possible answers to the question. We learn from a variety of voices and in a variety of modalities (e.g., verbal, visual, kinesthetic), so each lesson invites sharing from students and includes activities that engage various modalities. We remember what we believe will make us safe or happy or successful, so each lesson concludes with time for reflection. While time constraints in the classroom tempt teachers to skip reflection, it would be like eating without digesting. We all benefit from time with questions about how what we just learned could be useful in our lives. If we recognize how the newly acquired knowledge will somehow make our lives better, we will remember it. And if it has no possible use, we will be unlikely to retain it past the test, if that long.

Again, by way of introduction, we have gone around and around what literacy might look like. But, having circled in, we are now ready to make it tangible. While you may have agreed with the points made so far about competency, literacy, context, critical thinking, enduring answers, stories, multiple voices, learning modalities, and reflection, the proof of the pudding is in the eating. In the next three chapters, we will explore the nine questions within Identity, Change, and Evaluation of nanotechnology. To guard against ungrounded generalities, we will use the microbivore as a touchstone, testing the nanotechnology patterns we offer for utility. Since a desire to learn may be a trait common to our readers, we offer this analysis of a fantastic medical tool as a bonus.

References*

1. Robert A. Freitas, Jr., Microbivores: Artificial mechanical phagocytes using digest and discharge protocol, *Journal of Evolution and Technology,* 14, 55–106, April 2005, permalink http://www.jetpress.org/volume14/freitas.pdf
2. Chris Toumey, Nano hyperbole and lessons from earlier technologies, *Nanotechnology Law & Business,* Vol 1.4, 2004.
3. Kofi Annan, Science for all nations, *Science,* Vol. 303, 13 February 2004.

* All links active as of August 2010.

6

How Do We Teach Nanotechnology's Identity?

> In addition to tools and devices, we should include systems and methods of organization. Any collection of processes that together make up a new way to magnify our power or facilitate the performance of some task can be understood as a technology.
>
> **Al Gore**

Do you already have microbivores surfing your bloodstream? As far as we know, they have not yet been invented, just imagined, so probably not. But how could you tell if you did have them? They would be too small to see. When, and if, they are invented, how would you get them? Perhaps you would get a trillion in a small vial, administered by your doctor. How would they be updated to recognize the latest virus or bacteria circling the globe and filling the headlines? Perhaps with a download similar to updating the virus software on your computer. How would they be removed from your bloodstream, once given a chance to do their job? How have they been tested? Who pays for the treatment? How does the manufacturer insure against failure? How does the organization offering virus and bacteria definition updates insure? How are they disposed of or recycled? What laws apply? Who funded development? Who owns the intellectual property? Who set standards for how they should behave—at least so that they are interoperable with surrounding equipment from various manufacturers ("I'm sorry, but your microbivores are incompatible with our microbivore evacuation system.")?

What Is Nanotechnology?

Infrastructure. Policies. Systems. No technology stands alone—and certainly no microbivore. They function in a context of devices and information. If we breeze past the question *What is nanotechnology?* assuming it quite obvious that the thing, the artifact, the tangible product is the answer, we will miss most of the technology, just as we would miss most of the iceberg if we focus just on the bit above the waterline. To appreciate this, imagine taking one of your artifacts, maybe a cell phone, back in time a few centuries. Without the cell towers, the communication backbone, the electrical grid, and a second cell phone, you have a paperweight. Until the battery runs out, it would serve as a feeble flashlight and passing novelty for those wondering how to test whether you are a witch.

As an activity in this lesson, younger students can select a technology they use and then draw its connections to everything it relies on. They will be identifying dependencies on other technology, networks, and ways we do things (protocols, both technical and social). Imagine a car with its connections to streets, parking lots, stop signs, traffic lights, speed limits, lane dividers, and center lines. Think of the gas stations, repair shops, factories, rubber plants, metal foundries, and silicon microchip foundries. Consider laws (e.g., seat belt, cell phone, texting), liability insurance, financing companies, car dealers, third-party evaluators (e.g., *Consumer Reports*), advertising agencies, and the media.

Older students can imagine a nanotechnology and then draw and write about the connections they anticipate. Integrating with social studies curriculum, students could pick an era to which they would transport back with a technology. Then, write or draw how it would fail or what other technology they would also bring back to keep it useful. Think about energy, information, and repair.

One middle school teacher using ICE-9 to teach technological (not specifically nanotechnological) literacy opens his discussion of the first question, what is technology, with a wager. He picks one of the smallest students in class and asks who thinks that student can lift the teacher (who weighs 100+ kg) off the floor and keep him aloft for a minute. At stake is extra play time wagered against spending part of recess studying. The teacher generously warns the students that he would not make a bet he was not sure to win, but he still gets takers. Then he produces a long, strong plank of wood, places it asymmetrically over a small block to form a lever, and stands on the short end. Students quickly connect the idea that even a lever is technology when they see the small student step on the long end of the plank to gently lift the teacher from the floor. Some students have "skin in the game" and long remember that one answer to "what is technology?" It is something that extends your abilities.

As a definition, "technology is applied science" or "nanotechnology is applied nanoscience" works so long as your definition of science is rather broad. The incandescent light bulb, icon for both idea and technology, came about in part because Thomas Edison tested hundreds of different filaments. Science is generally taken to mean proposing and testing predictive models, which would be quite generous as a description of Edison's brute-force repetitive testing. Though science is immensely important for the development of technology, scientists and engineers still come across phenomena they cannot explain, and sometimes build on it to make useful technology.

Also, the predictive models of science have limitations. Theoretically, the standard model of particle physics can explain the interaction of everything but gravity (and the newly postulated dark matter and dark energy). However, the mathematical complexity grows wildly beyond practical solution when the standard model is applied to much more than a single proton and single electron. Even nanotechnology consists of far more than a single proton and single electron, so we use approximate models to predict nanoscale behavior. And engineering of nanotechnology relies also on results from Edison-style trial-and-error explorations. Nanoscience is crucial to nanotechnology, but it is misleading to define nanotechnology as "applied nanoscience."

Understanding "what is technology" is important to understanding "what is nanotechnology," but what about the "nano"? We opened the first chapter with definitions of nanotechnology, one of which focused on deriving function from features sized in the 1- to 100-nanometer range. We emphasized function because just about everything made out of atoms—not just nanotechnology—has features in this range. Diesel soot has carbon features in this range, but unless its function is predicated on those nanoscale features, it is not, by this definition, nanotechnology. Students want and need a more tangible definition than scale range, so let us drill down to specific classifications of nanotechnology. Even older, more established classifications, like the taxonomy of life (kingdom-to-species), are subject to debate, so it will be no surprise that the following categories of "generations of nanotechnology" are not universally accepted. But the human brain loves to classify, and categories help us understand, so we forge ahead with what we have:

1. The first generation of nanotechnology comprised nanopowders and nanomaterials. Sunscreen with nanoscale particles that reflect ultraviolet light, but not visible light, does not require precise placement of atoms or even molecules. Nor do pants with fibers so narrow in width and depth (but not in height from the material foundation) that liquid beads and rolls off. How do these work? Well, that is question four, the last of our identity questions. Thank you for asking and giving us something to anticipate.
2. The second generation of nanotechnology is characterized by devices with molecular precision. Integrated circuits with line widths of just 32 nanometers and thickness of just a few nanometers rely on precise placement down to a few molecules. Of course, this is changing rapidly, and by the time you read this book, line widths may be half that. This takes us to question 5, how does nanotechnology change, which we will explore in the next chapter on Change.
3. The third generation of nanotechnology animates the molecular precision of the second generation. Machines—gears, rods, levers, hinges, valves—reduced to the nanoscale require even finer precision to make everything fit. Friction, an unavoidable side-effect at the macroscale, means parts are breaking at the nanoscale. The third dimension assumes greater importance with this third generation.

The microbivore includes nanoscale machines, but as a whole, is a fifth-generation nanotechnology, as we will see two categories on.

4. The first three generations of nanotechnology have been built with specialized tools and factories. The fourth generation of nanotechnology will be (it does not yet exist) a factory operating at the nanoscale so that objects can be assembled from generic molecular building blocks. Life assembles molecules into larger structures, like cells, trees, and whales. The concept of the matter compiler is of a nanoscale factory able to assemble nearly anything we want from a stock of suitable molecules. It is one logical step after the third generation nanoscale machines. Being able to assemble nearly anything for which you have a design places far more value on intellectual property than material objects. This new balance is already familiar, with software far more valuable than the discs on which it is delivered, but the matter compiler would take it much farther. If the shoe or car material is free but their designs come dear, our economy would transform—but that's part of question six, "How does nanotechnology change us?" Ah, so much to anticipate!
5. The fifth generation of nanotechnology will be the nanobot, of which the microbivore is our poster child. It may be more complex than the matter compiler, because it must operate in a variety of environments. The matter compiler can keep its environment clean and predictable, like a Lego™ assembly line with molecules instead of plastic blocks. The nanobot is useful because it operates in the "real world." The nanobot may also rely on the matter compiler, because a vast array of nanoscale manipulators might be necessary to assemble not just one amazingly intricate nanobot, but trillions of them. Our bloodstream is a vast network for a microbivore, so even billions might not be able to patrol it in a reasonable amount of time, or take on the hordes of self-replicating pathogens they hunt.
6. Perhaps the sixth generation of nanotechnological growth will be the self-replicating nanobot. Taking a hint from bacteria, which need no central factory, nanobots might be programmed to make copies of themselves until a sufficient force is assembled for the desired task. Then they switch from making more of themselves to hunting bacteria, scavenging carbon from the atmosphere, or building spaceships. As Mickey Mouse discovered in "The Sorcerer's Apprentice," self-replication can get out of control. Economics may tempt us to develop self-replicating nanobots, but fear of disaster—even our own extinction—could discourage us. But again we get ahead of our story, asking, What are nanotechnology's costs and benefits? How fortunate that we have a framework to organize what we want to know about nanotechnology!

Images and diagrams, like Figure 1.1, *The Scale of Things,* in Chapter 1, help students understand what nanotechnology is. Online simulations are more interactive. NanoMission™ offers free videogame-style software with which students can explore design of a nanoscale probe injected into the human bloodstream to seek out cancerous cells (a variant on the microbivore). We describe NanoMission later in this chapter. Online use of nanoscale probes (e.g., atomic force microscope) can be scheduled for class time. A few schools are fortunate enough to have access to the nanoscale probes themselves. Simulation software allows students in any school with computers to assemble atoms into structures and simulate their response to forces, linear and rotational, while recording forces and temperature. High school students using NanoEngineer-1, a free molecular simulation environment,[1] enjoy ratcheting up applied force until complex molecules explode, with atom-to-atom bonds changing color as they stretch. Simple machines can be assembled and animated in the simulator (actual fabrication is nearing possibility, but is still slow and expensive).

An activity—really an educational game—created for this first question in technological literacy could be adapted for nanotechnological literacy. In *Technology Gameshow* one student is shown a hidden technology (e.g., compact disc, cell phone, pencil, light bulb, computer mouse, watch), and the rest of the class works in groups to ask yes-or-no questions to figure out what it is. Beyond curiosity, a motivation for being the first team to figure out the hidden technology is being able to send one of your members to view the next technology, and then be the one answering questions. Teams discuss what they have learned from the answers to previous questions about the hidden technology and use consensus to determine their next question.

Some students look for patterns in which questions are the most effective. The teacher encourages this search for patterns by instructing a student scribe to write each question and answer in one of three unmarked columns. The teacher decides the column based on whether the question asks why we use the technology, how it works, or neither. After several rounds, each with a new hidden technology, discussion centers on the pattern of the three columns. Questions in the first column (why do we use it?) tend to be the most effective, in the second column (how does it work?) nearly as effective, and in the third column (neither) rather unenlightening. Students eventually guess the meaning of column one and column two, but the third "catch-all" column is, not surprisingly, hard, so teachers usually reveal its identity once students get the first two.

This activity has the excitement of a game, gives ample attention to the student answering questions (possibly selected by the teacher because their attention-seeking had been disruptive), and lets the teacher play facilitator instead of lecturer. Using it for the question "what is technology?" allows for very familiar technology to be used. Adapting it for the question "what is nanotechnology?" would require an earlier introduction to nanotechnology so students would have some basis for their questions.

In open discussion, students get to share their thoughts on questions like, "In your learning how to ask good questions, which of your questions were most powerful?" and "What did you think technology was before this lesson? What do you think it is now?" Teachers can lecture through facts and concepts more rapidly than students can discuss them, but hearing something from a peer has a bigger impact. And students phrase concepts differently than adults do. Taking time in class for sharing questions like these consolidates the learning from an activity like *Technology Gameshow.*

After sharing discussion among the full class comes individual reflection. The prompt questions may be similar to those in sharing (e.g., "What do you think makes a question powerful?"), but the scope is broader and more personal. Students are not answering about the activity, but about what will be useful in their lives. While students are writing in their reflective journals, teachers get a chance to catch up on work, scheduling, correcting, filing, etc. They should not succumb to this temptation of open time. Instead, they should also journal reflectively. What worked well in this lesson? What would they change? What might they want to clarify before the next lesson? What learning might they take away to be more effective or happier in their lives? The teacher not only benefits from reflection, but affirms its importance and value. Like a parent that not only requires their child to wear a helmet while bicycling, but wears one, too, the teacher is modeling consistency.

Prompts for reflection include: How would understanding "what nanotechnology is" make your life better? What do you care about? What would you like to change? To create? Take time to reflect on possible connections between those things and nanotechnology. This is what teachers offer students: time to consider the value of what they have just learned.

A homework assignment turns *Technology Gameshow* upside down and encourages students to think about thinking. Each student identifies some technology, from a light bulb to a lunar lander, and then lists up to 10 questions to zero-in on the technology. Since the technology is known, the student also indicates the yes or no answer to each question. The activity goal is to make it appear that the questioner is very clever, but did not know the technology until the last question. Depending on the age of the students, it can be a stretch to create the illusion that they do not know the technology when crafting the questions. The learning goal is to apply what was learned about powerful questions so that the technology is rapidly isolated, from a most general question (e.g., "Is it used for communication?") down to the final question pinning down the technology's identity. With the most powerful questions you know, how few do you need to identify the technology?

To turn *Nanotechnology Gameshow* upside down requires student familiarity with one or more nanotechnologies. Since powerful questions in this activity tend to ask about why we use something or how it works, this activity may work better after the lessons on those two questions (ICE-9 questions #2 and #4). Whenever the activity is introduced, students will benefit from thinking about what is known and can be known about any given nanotechnology.

Just as technology changes into new forms, so will nanotechnology, and our minds must be nimble to engage the new.

Technological change has confused people throughout history. Consider how electricity transformed machines from the mechanical, which could be deciphered through careful visual inspection, to electronic, whose solid state (once we replaced the vacuum tubes) makes it inscrutable. And then electrical engineers and computer scientists, born into the new age, found integrated circuits, insect-size blocks of plastic with many metallic legs, to be perfectly familiar building blocks. Web sites brought another change in technology. Where was the device that Google, eBay, or Facebook introduced? Their innovations were not devices, but techniques implemented in software on Web sites, accessed by people who already had the devices of computer and Internet connectivity. The impact on how people search for information, conduct commerce, or communicate with friends has been dramatic. By thinking broadly about what nanotechnology is, students will not be tripped up or confused by transitions analogous to mechanical-to-electronic or device-to-technique. They probably had no trouble with these transitions because they were born after them, but technological change appears inexorable, so transitions will come in new guise to challenge today's students as they age. Before we explore "change" (the subject of the next chapter), we have three more questions to illuminate in "identity."

Why Do We Use Nanotechnology?

Before we get to nanotechnology in particular, let us consider why we use technology. One of the authors was guiding a classroom of 10-year-olds in a technological literacy activity called *Trees of Why* when he nearly intervened. Groups of three to five students used a simple consensus technique to decide what broad reason for using technology they would draw. The first student group to finish the previous activity took as their prize *entertainment* from the list we offered. They drew a tree with the trunk labeled "entertainment" and the branches showing more specific forms. They labeled the leaves with specific technologies: Xbox, PlayStation, TV, etc. The next group to finish selected any reason why we use technology from the list, except for the already-chosen *entertainment*. They took *communication,* imagining the leaves of cell phones, landlines, email, and maybe even pencil and paper.

The student group that the author nearly "helped," and is thankful he did not, chose *organization.* Since *exploration* with its tangible leaves of telescope, microscope, rocket ships, and laboratories was still available, he was tempted to push them to the easier topic. *Organization* could be abstract and frustrating for 10-year-olds. Abstract, maybe, but that student group seemed to find their challenge more interesting than the other groups did. Out of *organization* grew *land, everyday,* and *time. Everyday* sprouted *schools* and *streets. Streets* developed *tickets, dividers, meters, stoplights, parking places,* and *police,* as shown in Figure 6.1. The students thought creatively about how technology organizes.

Another classroom activity reveals the power of asking why we use technology. The teacher (or a guest visitor, if available) pretends to be a time traveler, asking about various technologies distributed to groups of students. These groups try to explain each technology in terms someone

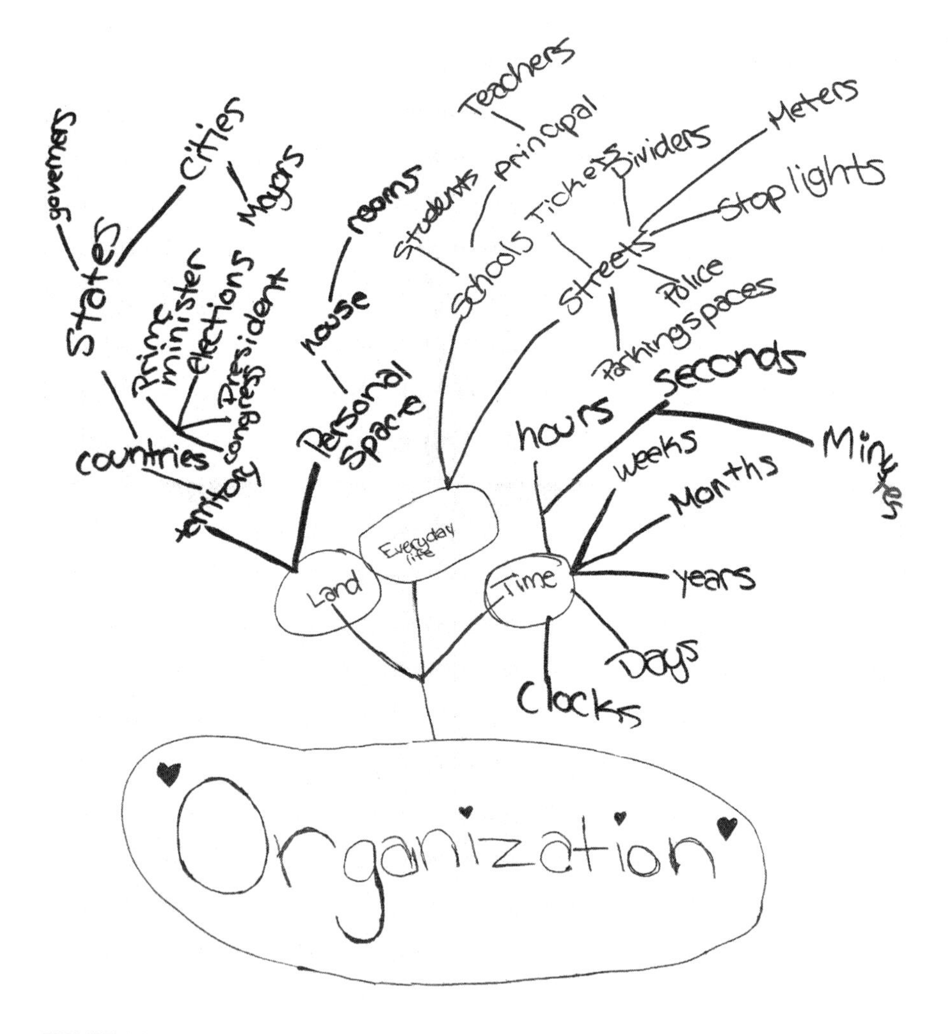

FIGURE 6.1
Trees of Why, as drawn by a student group of ten-year-olds for "organization." (Source: www.KnowledgeContext.org)

from 200 years ago would understand. A knowledgeable time traveler (teacher or visitor) may be able to teach a bit of history in the process of asking students questions. Students slowly recognize that attempts to explain a technology in terms of how it works tend to depend on other technologies, which need explaining in terms of still other technologies. However, explaining why we use it is more effective because those reasons change little over time. To encourage students to get this point, every student group listening to the one group explaining their technology is taking notes. Their worksheets have a row for each technology and a column for questions in the three categories: *why we use it* | *how it works* | *other.* These are the same column headings that remained secret during *Technology Gameshow.*

In open discussion, students get to share their thoughts on questions such as "What did you notice about explaining technology to someone who knows nothing about it?" and "What were the most effective ways to do it? What did not work?" Class brainstorming of reasons we use technology can create a long list. Some skill, often offered by students, is required to identify which reasons are more specific examples of other reasons. If the class is engaged, this discussion activity can create a hierarchy of reasons and a launching point for the *Trees of Why* activity described at the beginning of this section. We move from a class discussion of the big reasons we use technolopgy to a small group activity to draw specific examples that fit within those big reasons. This creates a balance of large/small group and verbal/visual modes of thinking. Some students may feel more comfortable sharing in small groups or in drawing.

Reflection questions ask "Did these activities change how you think about technology?" and "Tomorrow, if one of your friends showed you a completely new technology you had never seen before, what questions would you ask to understand it?" If that friend could present a microbivore, what questions would make it understandable? Ask why you would use a microbivore, and one answer is for health, more specifically medicine, and even more specifically infection. Unless the human is immune-compromised, a further specificity might be for emergent infections against which the human immune system has trouble creating antibodies.

The book *Technology Challenged*[2] (written by one of the authors of this book) offers a dozen major categories for why we use technology:

1. Food
2. Shelter
3. Communication
4. Transportation
5. Commerce
6. Art

7. Religion
8. Health
9. Entertainment
10. Organization
11. Conflict
12. Exploration

Biologists will be the first to note that we missed the biggest reason that life—not just human life— does anything: to procreate. Add that at position zero for 13 and we will also have the reasons we use nanotechnology. Recognizing these categories will make it easier to understand novel developments. Of course there is nothing unique or perfect about this list. Value comes in having an approach to making sense of a new technology. These categories organize information and guide our questions. A different set of categories for why we use nanotechnology is more specific and, in some situations, might be more useful:

1. Manipulate
2. Store
3. Sense
4. Communicate
5. Convert energy
6. Compute/decide
7. Protect (house/contain)

A microbivore satisfies all seven reasons. It *manipulates atoms* by grabbing pathogens and decomposing them into harmless pieces to be flushed away on the bloodstream. It *stores atoms* in the form of enzymes used to decompose pathogens. It senses particles it touches to identify those pathogens it has been programmed to consume. It *communicates* internally between sensors and manipulators to move an identified pathogen into its mouth, and externally when it is instructed to separate itself from the blood plasma in a centrifuge (it can adjust its buoyancy, and nonautonomous components in blood plasma cannot). It *converts energy* from its store of oxygen and glucose to mechanical movement of cables. It *computes* and *decides* what actions to take with manipulators, opening of orifices, and release of enzymes based on information from sensors. It *protects* its own mechanisms, energy, and enzymes inside a capsule designed to be inconspicuous to the human immune system.

The Foresight Institute, a nonprofit corporation that educates society about the benefits and risks of nanotechnology, offers six challenges.[3] They do not claim these categories are the only reasons we would use nanotechnology, just the most pressing for the good of humanity:

1. Providing renewable clean energy
2. Supplying clean water globally
3. Improving health and longevity
4. Healing and preserving the environment
5. Making information technology available to all
6. Enabling space development

While asking why you use something is a good first question when encountering a new technology, sometimes developers benefit from asking this, too. The microprocessor (now ubiquitous in computers, cars, telephones, kitchen appliances, and more) was, for a time, an orphan. Intel Corporation, initially a memory chip producer, developed the first microprocessor for Busicom, a manufacturer of handheld calculators. When the market price of calculators plunged, Busicom canceled the project. Convinced that the microprocessor would have value elsewhere, Intel purchased back the intellectual property from the contract and started pitching the microprocessor for many applications. At a conference in the New York Hilton, Intel's glowing description of the market for microprocessors as eventually reaching into the millions was questioned. Someone asked, "Where are they all going to go? It is not like you need a computer in every doorknob!"[4] Years later, that same hotel replaced conventional, metal keys with electronic cards ... and every room's door contained a microprocessor. Perhaps promoters of the microprocessor generated their own lists of reasons to use technology.

Where Does Nanotechnology Come From?

The story of the nearly orphaned microprocessor is acted out by middle school students studying "Where Does Technology Come From?" In lesson four of ICE-9, student groups choose or are assigned a story of invention in the form of a short reading. The group is to play act, creating their own informal script to convey the gist of the inventing process. One group's invention leads to another group's invention, which leads to the next, motivated because the preceding invention has some limitation addressed by its successor. The stories flow from the electric battery to electric generator, light bulb, and vacuum tube. The vacuum tube inspires the transistor, which leads to the integrated circuit, microprocessor, and personal computer.

Students learn that sources of invention vary. Accident leads to invention: a frog's severed leg jumps when touching two metals, leading to the *electric battery*. Necessity is the mother of invention: vacuum tubes burn out, so long-lived *transistors* are developed. Plan leads to invention: connecting tens of thousands of transistors takes too long and leads to too many mistakes, so *integrated circuits* make most of the connections with an automated template (recently exceeding a billion transistors on a single chip … and growing). Invention is the mother of necessity: *microprocessors* are invented and seek a need to satisfy. Sharing, students discuss:

- What technology was invented in a way other than combining two earlier technologies?
- Do you think technology today is invented the way we saw in the stories? Why?
- Does a new technology always improve on an old one? Think about video games, writing implements, lighting, and other categories of technology. Do some new versions have problems earlier forms did not?

Before students participate in this activity, *One Thread in History,* they try another, *Timeline.* For this activity, each student is given a piece of paper with the name of a technology and the year of its invention. The goal is to line up chronologically by discussing your technology, without revealing your year. Discussion of why one invention might precede another is key. Students naturally look for patterns. Does one invention include another and, so, must follow it? Does complexity suggest recent invention? Does knowledge of history simply place some inventions?

Timeline is a physical and personal activity. Friends, each associated with an invention, move and change places. Spatial memory complements verbal and visual memory. Students who have difficulty remembering what they read on a chalkboard or hear in a lecture may readily remember that their best friends, representing *windmill* and *airplane,* stood far apart, with *windmill* (800) far earlier than *airplane* (1903). That may cause them to wonder why some technologies were invented before others, and why some were not invented earlier. Some answers to such questions come in lesson 5 of ICE-9: Why does technology change?

As a matter of logistics, teachers find that a circle of students, not a straight line, makes it easier to communicate with students and for students to see each other. When the activity is done indoors, a circle (or approximation) is generally forced by the size of the room.

In a nanotechnology class for advanced high school students, one of the authors suggests five answers to where nanotechnology comes from:

1. Technology (physical and conceptual)
2. Dense population (physical and virtual)
3. Specialization and interdisciplinary work

4. Biomimicry
5. Design or evolution

We have a "fat finger" problem: human fingers are about 10 million nanometers thick. Nanotechnology comes from other technology because we cannot manipulate directly at the nanoscale. Richard Feynman suggested, as part of the thought experiments for which he became famous, that we could build small machines, which could build smaller machines, which could build smaller machines, and so on.[5] He was inspired by the pantograph, a machine that scaled down a drafter's tracing, before computer-aided design made scaling a trivial function. Technology like the atomic force microscope and variety of electron microscopes has led, and continues to lead, to nanotechnology.

Feynman offered no new physical technology, but his speech contained concepts that contributed to the development of nanotechnology. Conceptual is as important as physical when considering where technology comes from. Consider how slowly stone tools developed. They first appear in the archeological record 2.6 million years ago. While the exact date is subjective, depending on what one considers evidence for intentional crafting (and pending new archeological discoveries), it makes one wonder, what was going on?

> How could developing stone tools have been so slow? Were humans that much less intelligent back then? In answer to that question, consider the island of Tasmania and an old TV show, *MacGyver.* As anywhere, Tasmania suffered from occasional famine. Unusual, however, was that for about 4,000 years Tasmanians, surrounded by rich oceans, did not fish. They did not think of fish as "food," much as most Americans and Europeans rarely think of insects as food, even though people in many parts of the world recognize them as highly nutritious.
>
> Now, on to *MacGyver,* the television show about a resourceful hero who uses a paper clip to short-out a nuclear missile, a chocolate bar to plug an acid leak, and a cold capsule to trigger a homemade bomb. What did he have that most people with easy access to paper clips, chocolate bars, and cold capsules lack—other than life-threatening situations every week? Information. Most of us lack information about missiles, acid, and bombs just as the Tasmanians lacked information about fish. Similarly, before we invented stone tools, we lacked information about stone tools—a seed for technology.
>
> Our vast interrelated network of technology is like a crystal. The molecular components of crystals can float around in liquid (non-crystalline) form until they come in contact with a "seed" crystal. This seed is literally a few molecules that have already been stacked into a crystal structure. These cause more molecules to come out of solution, adding themselves to the structure. Integrated circuits are made from a giant silicon crystal grown from a tiny silicon seed. Stone tools may have been the conceptual seed from which all technology since has grown.[6]

Information is what Feynman brought us with his "There's Plenty of Room at the Bottom" speech.[5] Information is also the answer offered by Eric Drexler in his books *Engines of Creation*[7] and *Nanosystems*.[8] The fictional character MacGyver might have said, "It's not nanotechnology until you conceive it." Ideas build on tools (e.g., electron microscopes, nanomanipulators, nanotweezers) and on processes (e.g., plasma arcing, vapor deposition, ball milling, self-assembly, and nano-lithography).

The archeological story of Tasmania includes the loss of technology. Tasmanians had fishing hooks before they did not. Their neighbors on Australia had them and did not lose them. How do you forget a technology? Jared Diamond speculates that loss can come from insufficient population.[9] With approximately 4,000 inhabitants, Tasmania could miss an occasional baton-exchange, to borrow a term from track-and-field sports. Australia, with 300,000, would always have some group that continued to use and teach the use of technology already developed. In Australia, any group that lost the last crafter of a technology might either copy a group that retained it, or be marginalized and replaced by that group (if the technology were critical and coveted). Technology was apparently not lost on Australia.

Insufficient human population is no near-term threat on earth. Quite the reverse: overpopulation threatens our sustainability, and nanotechnology may be called on to help. But nanotechnology has developed in part because, on the path to overpopulation, increasing population density has made it easier to exchange tools and ideas (which sometimes allows greater productivity and denser population, thereby feeding the process). Development of nanotechnology relies on exchange of information, which can come from density of either physical or virtual populations. Universities, research laboratories, and regions (e.g., Silicon Valley) offer physical concentrations of brains, ready to feed off each other. The Web offers virtual concentrations of focused information (Foresight Institute, nanoSig, and Nanotechnology Now).

Specialization and interdisciplinary work also create new nanotechnology. This sounds contradictory, but advances in nanotechnology come both from specialization in a field (e.g., in physics, surface chemistry, biochemistry, genetic engineering, computer science, electrical engineering, mechanical engineering, materials science) and from connecting two or more fields. The microbivore illustrates an intersection of multiple fields. Although K–12 teachers can discuss these sources of nanotechnology, few will be educated as a specialist in one of these fields, and fewer will have experience in crossing disciplines in order to innovate. Videos of people who are both specialists and explorers across disciplines are, therefore, very useful in class. One video presentation available online is a TED Talk on "biomutualism."[10] The energetic talk, itself supported with video of geckos and robotic geckos, illustrates specialization, crossing disciplines, and our next answer to where technology comes from: biomimicry.

Biomimicry, adapting designs from life, leverages billions of years of Nature exploring the possibility space for what works at the nanoscale. Gecko feet show how to stick to a surface using just the attractive force present between atoms closer than 2 nm.[11] Spider silk demonstrates very high tensile strength at lengths we have yet to reach with carbon nanotubes. Abalone shells arrange "chalk" into resilient, impact-dispersing sheets. Algae photosynthesize and capture CO_2. Able to land on a bacterium and inject molecules, viruses may be a proof-of-concept for nanobots. Cells are good molecular assemblers, with the ribosome in the cell building proteins according to genetic instructions. DNA can be designed to fold and attach to itself, allowing us to create "DNA origami" that will form a variety of nanoscale shapes. With nanotechnology, we usually wish to do something different than is already done in Nature, so we find life that does something similar, or does a few things that could be combined or sequenced toward our design.

> Many of the most interesting problems in computer science, nanotechnology, and synthetic biology require the construction of complex systems. But how would we build a really complex system ...?
>
> **Steve Jurvetson**

Venture capitalist Steve Jurvetson has studied nanotechnology and complex systems. He suggests that very complex systems cannot be *designed*, but could be *evolved*.[12] He contrasts these two approaches to creation:

Designed Systems

- Predictable
- Efficient
- Controllable
- Subsystems easily understood, allowing reuse
- Break easily
- Proven on only simple problems

Evolved Systems

- Directable
- Adaptive
- Subsystems inscrutable
- Robust
- Resilient
- Proven on complex problems

While most technology is still designed, guided evolution has been used for antenna design. Physically building and testing devices is slow, so this approach tests candidate structures inside a computer simulation. The least promising structures are culled, replaced by mutations of the most promising structure (plus an unmutated form, just in case that is already near-optimal). The faster computers get, the more generations and mutations can be simulated. If matter compilers ever make fabrication sufficiently fast and cheap, guided evolution may step outside the computer simulation to a computer-controlled real environment. But the principles of evolution will persist: variation, selection, and retention. Mutations and combinations of structures create new variations. Environment selects some of those structures as more fit. Information specifying those more fit is retained for specifying the next generation.

Students learn about where technology or nanotechnology comes from on several levels of abstraction. Lining up by the invention year of a technology written on a sheet of paper, as in *Timeline,* is the most straightforward. Acting out skits of invention, as in *One Thread in History,* raises the level of thinking to consider human motivation. They must consider the environments conducive to invention when exploring the answers technology (physical and conceptual), dense population (physical and virtual), specialization and interdisciplinary work, biomimicry, design or evolution. These answers to where nanotechnology comes from also connect to social studies. This is important context, leading to science and engineering answers to the fourth of the ICE-9 questions…

How Does Nanotechnology Work?

Many equate understanding something with knowing how it works. Those students who pursue studies in nanoscience will probably be drawn to this exploration of science and engineering. The twin challenges that thread this section are most serious here:

1. What can K–12 students understand? Details of how nanotechnology works rests on advanced science, math, and engineering beyond all but the prodigies.
2. What understanding will endure in value, applying to future technology, as well? What patterns will illuminate current nanotechnology, yet also apply to that which has yet to be invented?

In middle school classrooms two online activities (part of ICE-9 and, therefore, free of cost) teach the power of repeated components and the emergent properties. These aim at technological literacy; later, we will introduce an online activity specific to nanotechnology.

Game of Life, a web-based activity, starts on graph paper so that students learn the rules instead of simply clicking frantically while their eyes glaze over. Based on computer scientist John Horton Conway's cellular automaton,[13] each square on the graph paper can be empty or have a cell, suggestive of life, but simpler. The rules for determining the next generation are simple: a cell dies of loneliness if it has fewer than two neighbors (in the eight adjacent squares); a cell dies of overcrowding if it has more than three neighbors; a new cell is born in an empty square that has exactly three neighbors. The goals are to predict what will happen to a given configuration of cells and to devise configurations that have certain properties. Some configurations of cells die off quickly. Others oscillate between two states. A few travel, while others "bloom." Once students have calculated a few generations by hand on graph paper, the web-based version[14] comes to the rescue, allowing them to quickly step through generations.

The learning goal is to understand *emergent properties,* which can make behavior of complex systems hard or even impossible to predict. Student intuition may be that if the parts are simple, the whole will be ... unless they have studied complex systems like weather or ant colonies. Simple nanotechnology will often be deployed in complex systems (ecosystems or the human body), and nanotechnology will comprise complex systems, like microbivores. The sooner that *emergent properties* become intuitive to students, the more facility they will have when they either develop nanotechnology or evaluate its impact.

Mystery Boxes, a web-based activity,[15] invites students to click on virtual switches, observe virtual lights, and figure out the connection between the two. The enigmatic boxes between switches and lights are labeled with standard computer logic names (NOT, AND, OR, NAND, NOR, XOR) that underlie our digital systems. Analogous to Lego™ building blocks, these Mystery Boxes can be assembled into complex systems. Students play online with examples of antilock brakes and burglar alarm. On paper, students design their own system, learning that simple functions can be assembled into complex (some of which may exhibit emergent properties, learned about in the previous activity).

The *Mystery Boxes* activity is a conceptual bridge between a very simple system that most or all students can understand (e.g., if Switch A *and* Switch B are on, then the Light will be on) and more complex systems that may intimidate students (e.g., the logic of a burglar alarm that senses multiple windows and multiple doors, and has a deactivation switch, modeled in the activity). Analyzing a complex system by recognizing the readily understood components gives students confidence with yet more complex systems. With practice, students will eventually be confident in approaching even those systems that take much time and effort to analyze. Their approach has been proven successful again and again, so they are ready for the challenges they will encounter in university and beyond.

Students also see that complex systems can be described without specifying or understanding all the underlying details. Describing a burglar alarm's function without explaining how its components work is *functional abstraction,* an important concept in technology. It allows specialization in fields or technologies that are layered on other fields or technologies and serve as foundations for still others. Ecology does not reinvent biology, which does not reinvent chemistry or the underlying physics. Computer applications build on computer operating systems, which build on programming languages and the underlying hardware design.

Class discussion allows students to share their ideas on questions like these: What did you like most about the Game of Life? What was surprising? What was confusing? What did you like most about the Mystery Boxes? What was surprising? What was confusing? What strategy worked best for figuring out the Mystery Boxes? What things can you think of that are made up of other things, which in turn, are made up of still smaller things?

Reflection in individual journaling is prompted with questions like these. Did the pattern of simple things forming complex systems surprise you? Do you think it is real beyond the classroom activities? Do you think it works with things that are not technology? Have you ever seen anything happen that reminds you of this idea? And going farther: how might your day-to-day decisions be different if this pattern applies to all technology?

Both *Game of Life* and *Mystery Boxes* address general technological literacy. *NanoMission*™ addresses nanotechnological literacy with "a cutting-edge gaming experience which educates players about basic concepts in nanoscience through real world practical applications from microelectronics to drug delivery."[16] Modules can be downloaded from the Web for students to play in classrooms or at home. The four modules released are:

> *NanoImaging Module*
>
> Dr Neevil has created genetically modified algae which is manifesting in huge blooms turning lakes red, toxic, and fatal to humans and animals. Your mission is to identify this microorganism in order to develop a countermeasure and save the world.
>
> *Learning Scale Module*
>
> NanoMission scaling module enables you to visualize and understand the spatial relationships between objects at scales from the picometers through nanometers all the way up to gigameters.
>
> *NanoMedicine V1 Module*
>
> Join Dr. Goodlove and Lisa in the demo of the first module, nanomedicine, select a suitable vehicle to deliver an anticancer compound, and then navigate through the bloodstream to the site of the tumor, avoiding the body's natural defense mechanisms.
>
> *NanoMedicine V2 Module*
>
> This module enables you to get a better understanding of the processes involved in creating nanomedicine. You assume the role of a biomedical scientist aiming to cure cancer through observation and experimentation by building nanoscopic particles and measuring their effects on the patient at the cellular level.

The game modules have been recommended as an outsource for K–12 students by the Nanocenter in Albany, New York, and are promoted widely on YouTube by students. Check the Web site often for future releases.

In working with high school students, it is possible to bring out additional answers to how nanotechnology works, beyond emergent properties (Game of Life) or repetition and layers (Mystery Boxes). Three we consider are scale, light, and energy.

Movies with giant insects rampaging through cities were popular once, perhaps fueled by the remarkable strength of insects. Ants carry many times their weight, so it is easy to assume that an ant the size of a car would be a fearsome opponent. But scale is a funny thing, and an ant the size of a car would die. First, their legs would break because limb strength scales with area, but the load they carry scales with volume. Enlarge a 1-cm ant to 10 cm, and its legs will be approximately 10 × 10 = 100 times as strong. Unfortunately, it will weigh 10 × 10 × 10 = 1000 times as much. It would be like multiplying your weight by 10 without increasing your strength. You would be lucky to lie down and breathe.

Speaking of breathing, ants do not need lungs because oxygen diffuses from tracheae, simple tubes that at most compress slightly to encourage diffusion. The time it takes for diffusion increases with the square of distance, which means something that might diffuse across 100 nm in less than one millionth of a second would take 12 seconds to diffuse across 1 mm and 35 hours to diffuse across 10 cm. Relying on diffusion to oxygenate their cells, giant ants would asphyxiate.

So giant ants would break their legs and suffocate, but scale is not all carnage. Scaling materials from the macroscale down to the nanoscale greatly increases the surface area relative to the volume. That results in chemical reactions occurring much faster. This is explored in *Size Matters: Introduction to Nanoscience*,[17] a lesson developed by SRI International and the National Science Foundation (NSF) for high school students. It explores "concepts related to the size and scale, unusual properties of the nanoscale, tools of the nanosciences, and example applications" in five lessons. It is part of a four-unit program called NanoSense.[18] The other three units are:

1. Clear Sunscreen: How Light Interacts with Matter
2. Clean Energy: Converting Light into Electricity
3. Fine Filters: Filtering Solutions for Clean Water

Clear Sunscreen: How Light Interacts with Matter[19] explores the physics and chemistry of light in five lessons. Since nanotechnology features are smaller than the wavelength of visible light (400 to 700 nm), interaction with light is one way nanotechnology works. In sunscreen, particles can be sized to absorb the ultraviolet light dangerous to our skin (with wavelength in the 250 to 350 nm range), but not visible light. The white sunscreen made famous by lifeguard noses contains a wide variety of particle sizes, which absorb a broad spectrum of light.

Energy is key for active nanotechnology (passive nanotechnology, like pants that shed liquid, require no additional energy source beyond shaking your leg to agitate the droplets off). The microbivore illustrates the energy challenges of complex nanosystems. In general, complex nanosystems need sources of energy, whether chemical, solar, thermal, radioactive, or kinetic. They need to store energy, whether chemically (batteries, hydrogen tanks), electrically (capacitors), or kinetically (flywheels). They need to convert energy (photovoltaic effect, thermal differential, chemical)—addressed in the *NanoSense* module *Clean Energy*. And they need to handle waste energy, heat (entropy management).

Students exploring how nanotechnology works in K–12 will learn about many areas of study that might intrigue them, even if they do not pursue study in nanotechnology. If they do choose that path, this introduction will give them context for directing their studies. They will not—cannot—learn all the answers in high school, or even in university. Leading researchers in a variety of fields are constantly advancing human knowledge of how nanotechnology works.

What we can give them in K–12 are those principles about scale, interaction with light, energy, emergent properties, repetition, and functional abstraction that they can grasp and that will apply to whatever they encounter in the future. We can offer not just declarative knowledge (which may soon be forgotten), but intuition about relationships and awareness that many fields of study exist. We can share the awe that we feel when we survey the panorama of research, discovery, and development at the nanoscale. We can invite students to learn more about the science, math, and engineering that either excites them directly or that would enable them to explore the nanotechnology that does excite them.

Gaining a contextual understanding for how it works adds to the context of nanotechnology's identity. Understanding what nanotechnology is, why we use it, and where it comes from rounds out this foundation. It prepares students to explore change, which is one reason that learning about nanotechnology is pressing. Were nanotechnology a static field, much the same for the parents and grandparents of students, there would be much less urgency. In the next chapter, we look at change in nanotechnology and emerging technology in general: How does it change? How does it change us? And how do we change it?

References*

1. NanoEngineer-1 is available from Nanorex, http://nanoengineer-1.com/content/

* All links active as of August 2010.

2. Miguel F. Aznar, *Technology Challenged,* KnowledgeContext, Santa Cruz, 2005.
3. Foresight Institute, http://www.foresight.org/challenges/index.html
4. Speech by Dr. Joseph Bordogna, Deputy Director and Chief Operating Officer of the National Science Foundation, at the ABET Annual Meeting, October 28, 1999, http://www.nsf.gov/news/speeches/bordogna/jb991028_abet.htm
5. Speech by Richard Feynman, "Plenty of Room at the Bottom" December 1959, http://www.its.caltech.edu/~feynman/plenty.html
6. Miguel F. Aznar, *Technology Challenged,* KnowledgeContext, Santa Cruz, 2005, p. 41.
7. K. Eric Drexler, *Engines of Creation,* Anchor Books, New York, 1987.
8. K. Eric Drexler, *Nanosystems: Molecular Machinery, Manufacturing, and Computation,* Wiley Interscience, New York, 1992.
9. Jared Diamond, *Guns, Germs, and Steel,* W.W. Norton & Co., New York, 1999, p. 313.
10. Robert Full, Learning from the gecko's tail, 2009, http://www.ted.com/talks/robert_full_learning_from_the_gecko_s_tail.html
11. Gabor Hornyak, *Introduction to Nanoscience,* CRC Press, Boca Raton, 2008, p.736.
12. Steve Jurvetson, blog, http://jurvetson.blogspot.com
13. Conway's Game of Life, http://en.wikipedia.org/wiki/Conway%27s_Game_of_Life
14. Online Game of Life, http://knowledgecontext.org/Activities/life/index.htm.
15. Online Mystery Boxes, http://knowledgecontext.org/Activities/mystery-boxes/index.htm
16. NanoMission, Learning Nanotechnology through Games, http://www.nano-mission.org/index.php
17. Size Matters: Introduction to Nanoscience, http://nanosense.org/activities/sizematters/index.html
18. NanoSense project (2004–2008), http://nanosense.org/
19. Clear Sunscreen: How Light Interacts with Matter, http://nanosense.org/activities/clearsunscreen/index.html

7

How Do We Teach about Change in Nanotechnology?

> Change is the law of life. And those who look only to the past or present are certain to miss the future.
>
> **John F. Kennedy**

Change takes time, so the young do not easily recognize change has taken place. Students in K–12 were born into a world that may disorient many adults. In less than two decades, the Web moved from the domain of university scientists sharing data to becoming a necessity for information gathering, communicating, socializing, and even working. With connectivity available in coffee shops and libraries, and widespread use of smart phones, technology can be an extension of our consciousness. Technology moves many jobs to countries with cheap labor and then, even more quickly, automates processes into computers that can be located anywhere.

All this is normal for young people because they have lived with nothing else. But they are starting to experience the new because some technology—particularly information-based technology—changes at an accelerating pace. Inventor Ray Kurzweil studies and writes[1] about exponential change, finding it in:

- Computing power cost
- DNA sequencing cost
- Growth in Genbank
- Semiconductor memory cost
- Magnetic data storage cost
- Wireless bandwidth cost
- Internet hosts
- Internet data traffic
- Mechanical device size
- Nanotech science citations
- U.S. nano-related patents

Exponential growth is powerful. When something doubles (or halves) every few years, change may start slowly, but quickly becomes transformative. So young people start to notice—and be affected by—technological change not only because they are getting older, but also because it is accelerating.

So, how do we teach about change in nanotechnology and in technology in general? We start by identifying three aspects of it. *Technology changes.* Electronics and mechanisms shrink from thousands to hundreds, and now tens of nanometers. *Technology changes us.* Three-dimensional printers are proving the concept for a future of matter compilers that would change how we work and consume. *We change technology.* If there is ever to be a microbivore, it will be devised, produced, and maybe prohibited by people who understand nanotechnology. The next three sections explore how we teach change intrinsic to nanotechnology, its impact on us, and our roles for influencing it.

How Does Nanotechnology Change?

Middle school students get a sense of exponential growth from the activity *Calculate Time to Build a Microprocessor.* Early computers had about 18,000 switches, in the form of vacuum tubes. Current computers have billions of switches, in the form of transistors, integrated onto silicon chips with each transistor measuring just tens of nanometers. Exponential growth of the number of switches in computers enables fantastic new applications and prevents us from ever constructing computers the old fashioned way, connecting each switch by hand.

Students calculate how long it would take to connect a Pentium II microprocessor's 7,500,000 transistors if each connection took just a second. Of course there is much more to fabricating a microprocessor than just connecting a lot of transistors, but the magnitude of the result shows how much computing power has changed. Applying the concepts of units of measure, exponents, division, and significant figures, students calculate that it would take a year, working eight hours per day and five days per week.

One conclusion derived from class discussion is that microprocessors would not be common if they required that much effort to make. Students also connect back to the *One Thread in History* activity (see previous chapter), where the story of the invention of the integrated circuit described a problem called the *Tyranny of Numbers.* Motivation to create integrated circuits came from the expense, time, size, and unreliability of manually wiring together many transistors (imagine making no mistakes when wiring one transistor per second for a year). Analogous to the hand-copying of books being supplanted by printing presses, hand-wiring of transistors has been largely replaced by optical lithography for "printing" or etching patterns of transistors and their interconnects as integrated circuits.

One reason for students to personally perform the calculation in this activity is that this action has more impact than just hearing large numbers. In 2010, Intel started producing a microprocessor comprised of two billion transistors.[2] To many, "million" and "billion" are simply unimaginably large numbers. Only with time spent putting them into personal terms, do students begin to appreciate the impact. Updating the calculation to this new microprocessor shows that each one would require 267 years to assemble. Were we still using vacuum tubes as electronic switches, instead of transistors, the integrated circuit would not be possible, nor would processors of two billion switches. Class discussion invites connection of the concept of change to technology familiar to students:

- How have video games changed in your lifetime?
- Can you think of anything else that changes a lot over time? (e.g., cellular phones, computers)
- Why do you suppose it does this?

Taking the time to individually reflect in their journals, students are prompted with:

> If some technology could be either twice as powerful or half as expensive, which would you choose? Write an imaginary story about some technology that keeps doubling in power or halving in cost every few years. What is your world like when that technology is 100 times more powerful or 1/100 the price? (The technology could be anything: computer, car, TV, house, airplane, light bulb, x-ray machine, telescope, lever, clothing, stereo, flashlight, basketball shoes, etc.).[3]

In a nanotechnology class for high school students, exponential change is included in the fourth of five general ways that nanotechnology changes:

1. **Disappearing technology**: Nano-Tex™ pants are manufactured with fibers nanoscale in cross section so that a miniature forest of them keeps liquid droplets from adhering to the surface of the pants. The "nanotechnology" aspect of the pants is invisible, unless someone spills wine, juice, or water on them (when the hydrophobic effect from the scale of the fibers causes the liquid to bead and roll off). In the previous chapter, we described microprocessors disappearing into the doorknobs of hotel rooms. Nanotechnology will become inconspicuous not just because the scale of features is smaller than the wavelength of visible light, but because the functions it performs will become routine and accepted.
2. **Necessity's mother and daughter**: Quantum computing, just now moving from theoretical to practical, allows a computer to explore

many results simultaneously instead of sequentially, which is what conventional computers do. Desire for faster computation of certain kinds of problems that take a long time on conventional computers is one "necessity" that led to development of quantum computing. But working in the reverse direction is another phenomenon. As we make integrated circuit features smaller (to fit in more transistors), quantum physics predicts that electrons will be able to tunnel right through insulators. This "quantum tunneling" will lead to increasing errors and decreasing reliability if ignored. Designers of quantum computers have taken this "problem" of quantum effects and figured out how to make use of it. Quantum mechanics cuts both ways.

3. **Advantage, compatibility, risk, visibility**: New technology tends to be adopted if it (A) has an *advantage* over existing technology, (B) is *compatible* with existing technology in terms of manufacturing processes, distribution, and the ways consumers use it, (C) has low *risk* of financial or environmental failure, and (D) the advantages are clearly *visible.* Carbon nanotubes possess speed and heat-resistance advantages over silicon for conducting and switching electrons, but manufacturing requires different tools. Integrated circuit fabrication labs cost billions of dollars, so the financial risk of introducing a new technology is high. Because conventional silicon processes are highly developed and advancing, carbon nanotubes' advantages are not clearly visible.
4. **Autocatalysis**: Technology is used to create technology which creates new technology, resulting in exponential growth. This building on itself is an autocatalytic effect already seen with computers used to design and lay out the next generation of computers. Eventually, there will be a nanoscale assembler able to create nanoscale assemblers. That will be a powerful demonstration of autocatalysis.
5. **Memes**: Concepts or ideas may compete with each other, much as genes do in evolution by natural selection. Richard Dawkins coined the term "memes" for describing the competition of concepts in an ecosystem. Memes frame our thinking, as Feynman did when he described building with molecules. They will change nanotechnology by guiding what scientists and engineers think to explore. Biomimicry, the concept of copying what works in Nature is one meme changing nanotechnology. Another meme is the idea of nanoscale computing with mechanical components instead of electronic. Called "rod logic" because it implements a computer log with sliding rods and turning gears, it might seem a quaint throwback to nineteenth century computing, but scientists are showing that at the nanoscale, it can work better than electronics.

Advertisers pitch new technologies as being faster, having greater capacity, being lighter, being less expensive, looking better, or lasting longer (at least longer lasting on each charge of the battery, if not before you replace the whole thing with something newer). Thanks to advertising, students are aware of improvements in products. Teaching them some of the mechanisms underlying those improvements helps them think more critically about their selection and anticipate the nature of change to come. Those future changes will affect their health, work, perception of reality, and possibly even the fate of our species.

How Does Nanotechnology Change Us?

When we experience an electrical blackout, we appreciate how much electricity has changed us. Our homes become dark, refrigerators warm, computers inert, televisions blank, and rechargeable devices desperate. To study how technology changes us, students write an essay, *Day in the Life,* describing their day without any electricity, even that from batteries. The student handout lists many ways we use electricity. Students circle those they use and, as a class, brainstorm other ways before writing their individual essays.

For those who have trouble starting their story, the teacher offers, "Little did I know when I woke up that morning ..." After writing the essay, students form pairs to share with each other the benefit they would most miss with the loss of electricity. Volunteers may share with the class. For journal reflection we ask:

- Do you think about technology differently after imagining not having some of it?
- What do you imagine someone from the future might miss if they traveled back to our day?

As homework or extra credit, students can "read about an historical setting, perhaps from a social studies textbook, and write about what would happen if you transported a modern technology back to that time. How might it change the way the people in that ancient time work, communicate, play, or think about their world?"[4]

A nanotechnology class for high school students offers four answers to how nanotechnology changes us (or how it will). These can be developed in classroom discussion and expanded upon in groups and individual reports. While science, math, and engineering are important for understanding what is technically possible, discussion is still fruitful among K–12 students who have limited study in these areas. The scenarios are both personal and dramatic.

1. **Health and lifespan**: Nanomedicine in current and speculative form (e.g., microbivores) addresses health issues at the nanoscale. Human psychology suggests that, as we are further protected, many will take further risks, as already seen in correlations between riskier driving and the adoption of protective devices like airbags, antilock brakes, traction control, and even seat belts. But advances will not be completely offset. Improvements in water sanitation and availability of food more than doubled lifespan over a few centuries. Obesity and lack of exercise have only partially offset these advances.
2. **Work and recreation**: Valuable designs and nearly free manufacturing, as we predict with matter compilers, would automate manufacturing and assembly jobs that moved from developed countries to developing countries. There might be no wage low enough to make it worth waiting for a product to be shipped around the world, when it could pop out of a local matter compiler for little more than the cost of the design. Work requiring more knowledge and creativity may be mastered by computers made vastly more powerful by carbon nanotubes and molecular switches. What work will be left for humans? Exploration and recreation sound idyllic, but some governments and societies prefer the "market" to sort through those who are productive and those who are not. Students in K–12 will need to be intellectually nimble to figure out which skills will retain market value.
3. **Perception of reality**: Utility fog and virtual reality are where videogames may be headed. Virtual reality, with large goggles that offer a 3-dimensional view of a computer-generated world, do not require nanotechnology in their current, still primitive, form. Utility fog would dispense with the goggles and allow us to move in a space similar to the holodeck on the science fiction TV series *Star Trek: The Next Generation.* Microscopic nanobots (with nanoscopic features) numbering in the trillions or more would coordinate, linking arms, to simulate virtually any environment around the person. Another approach would be to place nanobots in the person's brain to stimulate senses in a coordinated manner, creating waking dreams of our own design. If those nanobots also stimulated dopamine and serotonin receptors in the brain, the effect could be much more addictive than drugs, given how real the experience would be. We might even lose track of what is real. Already, the Web has allowed a degree of that loss of reality, with people accessing sources of information that reinforce their beliefs to the exclusion of all dissent.

4. **Nature and survival of the species:** Runaway replication of nanobots could displace the human species. The free market economic system favors the most competitive, echoing survival of the fittest in ecosystems, meaning best able to procreate. Efficiency is a driver of technological development, and nanotechnology has the potential to be extremely efficient. In the case of nanobots capable of making copies of themselves, called the *gray goo scenario,*[5] they might be able to scavenge materials—effectively food—from anywhere: buildings, plants, or animals. Converting everything into copies of oneself is very efficient, but it has the side-effect of destroying all other life. Students in K–12 who think about sustainability may influence how nanotechnology is developed, assuring that efficiency is not the only driver.

Mythology and popular culture contain warnings about power getting away from its wielder (e.g., Frankenstein). Nanotechnology may turn philosophical thought experiments into ethical debates with serious, proximate consequences. Not all nanotechnology should be developed simply because it can be. In the previous chapter, we listed Foresight Institute's six challenges, directing nanotechnology toward the betterment of human society. Students in K–12 can discuss how they want nanotechnology to change us, and how they could influence it. The third of three questions about change focuses on just that: what roles we can play to guide nanotechnology.

How Do We Change Nanotechnology?

The stories from the activity *One Thread in History,* which helped to answer where technology comes from, return to help answer how we change technology. In the earlier activity using these stories, every student read every story. This time, students form groups and choose (or are assigned) one story to play act. They create their own informal script to convey the gist of the roles people play in the inventing process. One group's invention leads to another group's invention, which leads to the next, motivated because the preceding invention has some limitation addressed by its successor. The stories flow from the electric battery to electric generator, light bulb, and vacuum tube. The vacuum tube inspires the transistor, which leads to the integrated circuit, microprocessor, and personal computer. A script offered to teachers:

> Who remembers the stories we read about the frog and the battery? How about the generator, the light bulb, the vacuum tube, the transistor, the integrated circuit, and the PC? Today, we will get to perform skits to show how people in those stories changed technology. Don't worry: you won't have to memorize any lines. We will just act out the stories, and we can

> make up our own lines. You have seen the stories before. Your group will decide who gets which role. The skit (which may be videotaped) does not have to be long; it just has to help the audience understand what the technology was and what roles the people had in changing that technology.

While one group performs their skit, all other students take notes on a worksheet with a row for each skit and columns for:

- What was invented? Why?
- What were the people like?
- What motivated them?
- What made them successful?

Sharing in class discussion is described in the Teacher Guide:

> Teams share what main ideas they noticed about the technology that came before theirs. The first team does this for the last team, so every team has another to focus on. Pay special attention to: What was invented and why? What were the people like? What motivated the people? What made them successful? This will prepare students for Reflection.

For writing in their reflection journals, students are prompted with these questions:

- What problems do you notice in the world?
- What problem do you care most about?
- How would you personally like to influence technology in the future?
- What would motivate you to solve them? (for example, helping people, getting rich, figuring out something before anybody else, having a fun project, etc.)

K–12 students may feel far removed from influencing technology, much less nanotechnology, but there are many roles of influence they can play. They can discover, engineer, govern, promote, manage, invest, question, or teach, to list a few. To be effective in any of these areas, someone who would positively influence nanotechnology needs a contextual understanding (as ICE-9 offers) as well as a foundation in science, math, and engineering. Unfortunately many influence nanotechnology without any of this, which means they are doing so unaware of its connections, limitations, or possibilities.

It is not possible to know all the technical details, but it is possible to grasp enough of the context to know where one is ignorant. Knowing what we do not know is important. It tells us when to call on experts and may help us evaluate their counsel. Eric Drexler advised students to read widely of many scientific fields. Fields use their own terminology and their own unstated facts; common knowledge within the field may be quite uncommon outside it. Specialties within fields carry this to an extreme. Of course it makes sense to not repeat a specialty's common ground when communicating new advances, but for the outsider it is frustrating. Drexler suggests getting used to this feeling of understanding little of what one reads, because it allows one to get a sense of what is happening in a wide variety of fields.

Reading outside their familiarity is hard advice for K–12 students, who are rewarded for mastering the subject matter in each class. For them, not understanding is a warning sign of potential failure. If they can move beyond concern for grades for this activity, however, they can gain the panoramic view that will help them influence interdisciplinary fields like nanotechnology. The familiarity they gain from such challenging reading helps them make connections. Advances in one field might apply in another. Such clues can focus further learning.

Time is insufficient to be expert in many fields, so good communication skills facilitate communicating with experts in the fields one finds important. Ray Kurzweil makes this point:

> Technology-based innovation these days requires collaboration between different disciplines. This is because innovation today typically involves interdisciplinary work. So one thing that you need is experts in each of those areas.[6]

What we learn about how nanotechnology changes, how it changes us, and how we change it is our second layer of foundation for critical thinking about nanotechnology. Our first layer was about nanotechnology's identity: what is it, why do we use it, where does it come from, and how does it work? That supported our change layer, and together, they support the evaluation layer. We are ready for the next chapter, in which we ask what are nanotechnology's costs and benefits, and how do we evaluate it?

References*

1. Ray Kurzweil, *The Singularity is Near,* Penguin Group, New York, 2005, pp. 69–84.
2. World's First 2-Billion Transistor Microprocessor, http://www.intel.com/technology/architecture-silicon/2billion.htm
3. ICE-9 curriculum download, Teacher Guide, p. 34, http://knowledgecontext.org/Curriculum/download.htm
4. ICE-9 curriculum download, Teacher Guide, p. 39, http://knowledgecontext.org/Curriculum/download.htm
5. Gray goo scenario, http://en.wikipedia.org/wiki/Grey_goo
6. Ray Kurzweil, http://www.kurzweilai.net/meme/frame.html?main=/articles/art0563.html

* All links active as of August 2010.

8

How Do We Teach Evaluation of Nanotechnology?

> We haven't formulated and agreed upon a way of making good decisions about the powerful technologies we're so good at creating.
>
> **Howard Rheingold**

How do we decide if a proposed nanotechnology is good? If we prepare K–12 students to think critically, they will decide by first asking, good for whom? Themselves? Their community? The environment? They will reflect on their own values or the values of those who will be affected, because "good" is a value judgment. Then they will weigh the costs and benefits they figure out themselves or those proposed by experts. They will be able to weigh costs and benefits because they will have a contextual understanding of nanotechnology. Perhaps that will come from studying the questions we have covered in the previous two chapters:

1. What is it?
2. Why do we use it?
3. Where does it come from?
4. How does it work?
5. How does it change?
6. How does it change us?
7. How do we change it?

The foundation of these seven *identity* and *change* questions enables critical thinking about costs and benefits. In the next section we will look for answers to what nanotechnology's costs and benefits are. Then in the section following that, we bridge from the objective costs and benefits to subjective values. Two people might agree about a microbivore's costs and benefits, but disagree about whether it would be good or not. Suppose microbivores had a very high likelihood of saving the life of a loved one, but cost as much as simpler treatments for 100 people you do not know? Ethics are part of critical thinking.

The American Association for the Advancement of Sciences (AAAS) includes in their Project 2061 educational materials[1] these questions for evaluating a new technology:

- What are alternative ways to accomplish the same ends? What advantages and disadvantages are there to the alternatives? What trade-offs would be necessary between positive and negative side effects of each?
- Who are the main beneficiaries? Who will receive few or no benefits? Who will suffer as a result of the proposed new technology? How long will the benefits last? Will the technology have other applications? Whom will they benefit?
- What will the proposed new technology cost to build and operate? How does that compare to the cost of alternatives? Will people other than the beneficiaries have to bear the costs? Who should underwrite the development costs of a proposed new technology? How will the costs change over time? What will the social costs be?
- What risks are associated with the proposed new technology? What risks are associated with not using it? Who will be in greatest danger? What risk will the technology present to other species of life and to the environment? In the worst possible case, what trouble could it cause? Who would be held responsible? How could the trouble be undone or limited?
- What people, materials, tools, knowledge, and know-how will be needed to build, install, and operate the proposed new technology? Are they available? If not, how will they be obtained, and from where? What energy sources will be needed for construction or manufacture, and also for operation? What resources will be needed to maintain, update, and repair the new technology?
- What will be done to dispose safely of the new technology's waste materials? As it becomes obsolete or worn out, how will it be replaced? And finally, what will become of the material of which it was made and the people whose jobs depended on it?

Project 2061 offers many questions, so we will not ask them all if confronted with only a casual choice. But if we are developing, funding, approving, or questioning a powerful new nanotechnology, they are all important questions. Students can practice with ICE-9, applying those nine questions to technology they use, technology they wish they had, and technology that worries them. That practice will make the detailed questions from Project 2061 familiar, and perhaps even obvious, when a great deal hangs in the balance. So how do we teach K–12 students to think about costs and benefits?

What Are Nanotechnology's Costs and Benefits?

How do you measure cost? Many people think immediately of money, but time, health, safety, justice, environment, and sustainability are also ways we measure cost. Middle school students start their exploration of costs and benefits with a drawing activity called *Cost Bubbles.* Teachers introduce the ideas with this:

> The success strategy we will learn today is to compare the costs and benefits of technology. The cost of a console video game may be $200, but a benefit is not spending $0.25 per game for 1000 games at a video arcade ($0.25 × 1000 = $250). Costs and benefits may be measured in money, but there are other ways to measure them. Costs and benefits of a video game may be measured by:
>
> 1. Time (cost: you may spend all of your time playing with your video console)
> 2. Health (cost: you may not exercise, lose muscle, and get out of shape)
> 3. Jobs (benefit: you may learn how to program video games and get a job)
>
> Going beyond the video game example, there are other ways to measure costs and benefits:
>
> 1. Fairness (cost: does the Internet give an unfair advantage to those who can afford access?)
> 2. Sustainability (benefit: solar energy is sustainable because it does not consume limited resources)
> 3. Environment (cost: gas-powered cars pollute the air)
> 4. Advances in knowledge (benefit: the Internet allows people to share knowledge and ideas, creating even more knowledge)
> 5. Culture (cost: when MTV is broadcast in a third world country, does it destroy the local culture?)
> 6. Future benefits (costs and benefits: who can predict what nanotechnology will bring in the future?)
>
> Can anyone think of other ways to measure costs or benefits?

After brainstorming as a class, students select a familiar technology, such as a car, draw it in the middle of their paper, and surround it with bubbles. In each bubble, they write a different cost for their technology. It might look something like Figure 8.1.

When comfortable that students grasp the variety of ways we measure cost, invite students to add *benefit* bubbles to their technology. Drawings can be shared in class, so students see other perspectives and ideas. The next activity is *PMI Chart,* which has this objective:

> Practice analysis of technology using a simple strategy. Why is this important? Analyzing technology is basic to making educational, career, civic, and recreational decisions that are strongly influenced by technology. Technological factors are becoming more common in these decisions. An easy-to-use analytical tool can replace impulsive decisions or those based on someone else's opinion.

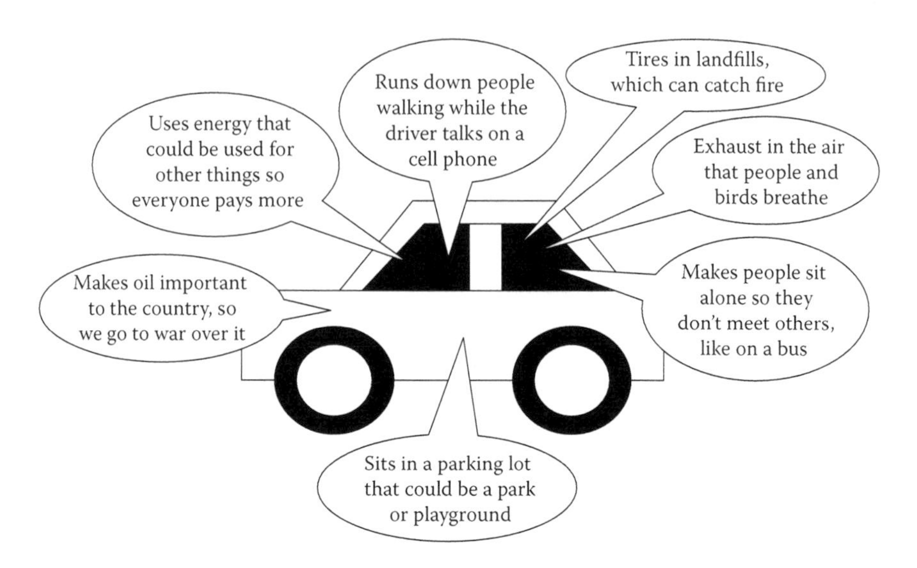

FIGURE 8.1
Cost bubbles for a car. (Source: www.KnowledgeContext.org)

For *PMI Chart,* students draw a table with three columns, headed *Plus, Minus,* and *Interesting.* They again select a technology that interests them, this time writing down several things about it that are positive or good (plus), several that are negative or bad (minus), and several that are simply interesting. This becomes raw material for a short essay, structured to make it easy for even those students who have trouble writing. The essay has four paragraphs. The first paragraph talks about what is interesting, the second talks about what is a plus, the third about what is a minus, and the fourth concludes whether the pluses or minuses seem most persuasive.

The first two activities were warm-ups, getting students comfortable thinking about costs and benefits. The next activity points to a pattern in costs and benefits: benefits are often direct and anticipated, while costs are often indirect and unanticipated. Consider a cell phone. The person using the cell phone directly benefits from its use. That is anticipated. If the user loses focus while driving and talking, someone else not using the phone may indirectly suffer the cost of being run over. That was unanticipated by the developer or seller of the cell phone. The motivation of selling a product focuses attention on the direct beneficiary (the potential buyer). Nobody wants anyone else to indirectly suffer costs from the product, but those unknown people will probably not get as much attention from the developer or the seller.

Three Dimensions of Technology's Impact has students draw a three-dimensional table similar to Figure 8.2.

Students again select a technology of interest and, perhaps in groups, think up costs and benefits. They may have experience or knowledge of a

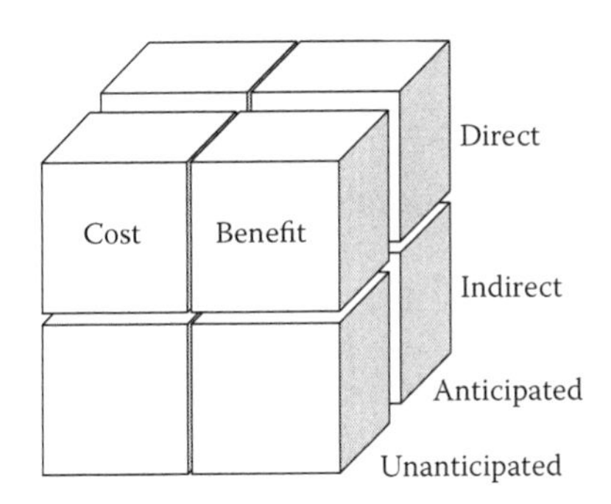

FIGURE 8.2
Three Dimensions of Technology's Impact. (Source: www.KnowledgeContext.org)

technology already used, so they will know of unanticipated consequences. If they imagine a new technology, like a microbivore, they will have to decide if some of the consequences they imagine might be unanticipated by others. Clearly, it is hard to write consequences in the four squares in the back. In fact, one square (anticipated indirect cost) is completely invisible in this perspective drawing. The drawing is a mental tool for keeping track of the dimensions. Consequences can be listed next to it.

This activity can work even better as a research project. Brainstorming may not raise most of the consequences, especially if the technology is not very familiar. One challenge of research projects is organization. Young people can find it particularly challenging to move from a blank sheet of paper to a report. And starting can be the hardest part. Using *Three Dimensions of Technology's Impact* gives them structure. The report could have 10 paragraphs: an introduction to the subject, one paragraph on each of the eight categories, and a summary paragraph pointing out surprises and offering a general conclusion.

What other patterns in costs and benefits can we teach? Some of the most interesting involve trade-offs between two ends, the more you get of one, the less you get of the other. New technology allows us to choose a new spot on the continuum, but rarely more of both (or less of both). Six trade-offs that we invite students to explore (and for which to seek examples or counterexamples) are Enabler vs. Crutch, Efficient vs. Resilient, Complex vs. Predictable, Catastrophic vs. Chronic, Control vs. Freedom, and Progress vs. Obsolescence.[2]

1. **Enabler vs. Crutch**: The more useful something is, the more dependent on it we become. Electricity is a familiar example. When the year 2000 approached, many worried that computer programs storing just the last two digits of the year would be confused by "00" and stop working. Though it turned out to be much less serious than headlines predicted, the "Y2K Bug" awoke the general population

to how dependent the world is on computers, including those embedded in our cars and buildings. Nanotechnology may prove so transformative, giving us nearly free products for instance, that we become completely dependent on it for at least some applications. Is it unavoidable that we become dependent on very useful technology? Short of not using it (who is going to give up their cell phone, much less electricity?), the alternative is to maintain backup systems and alternatives. But this is expensive and sometimes impractical. Would you have a backup refrigerator at home that does not depend on electricity?

2. **Efficient vs. Resilient**: Technology tends to trade resilience for efficiency, and that leaves us vulnerable. Large-scale agriculture is very efficient, so less than 2% of the developed world has to work as farmers. However, the system is brittle, with dependence on oil and monocultures vulnerable to pests. This echoes back to Chapter 6, where we described evolved systems as resilient. Life tends to evolve resilient systems over efficient systems. A highly efficient plant or animal that cannot survive an unusually cold winter or dry summer makes room for those that can. With future nanotechnology design, we may choose a different balance between efficiency and resilience.
3. **Complex vs. Predictable**: Who wants complex? Telephones and remote controls for audiovisual equipment can be crowded with features most users never learn. Cars break in new ways, now that they have computers control braking, traction, theft-deterrent, environment, and entertainment systems. We get complex technology because either we want some new feature (antilock brakes can be very useful) or those selling the technology need to introduce new features so they can convince us that the technology we have is no longer good enough for us. Sometimes complexity is just hard to avoid. Nuclear power plants have lots of systems to monitor and control. We can choose to make technology simple or complex, but if we choose the latter, we tend to get unpredictable. In part this comes from something we learned in Chapter 6: some kinds of complex systems have emergent properties, behavior of the system that is hard to predict based on behavior of the components. But even without emergent properties, complex means more components that could fail. And if we choose efficient over resilient, then failure of any component can cause system failure. There is not a right answer between complexity and predictability, but there can be awareness of the trade-offs we are making. Nanotechnology will allow us to make systems far more complex than those we currently have, so we should be conscious of when new systems are placed in mission-critical or life-and-death situations.

4. **Catastrophic vs. Chronic**: Dying of exposure to cold is sudden and catastrophic. Dying of lung disease from smoke particulates while staying warm next to a fire is slow and chronic. Dying in a house fire is sudden and catastrophic. Dying of asbestos inhalation in a fire-resistant house is slow and chronic. We can trade a catastrophic cost for a chronic one, or visa versa. This trade-off transcends technology, so it seems likely that nanotechnology will continue the pattern. Microbivores, as described, appear to offer just the benefit of avoiding death from a blood-borne infection. But, if they are developed, it will be critical to watch for unanticipated costs. If those costs are chronic and subtle, they will be hard to connect back to the microbivores. Humans already exhibit a diversity of chronic ailments we have yet to connect to causes. Are we sick because of an infection, the food we eat, the air we breathe, or the technology we are near? Ideally, the nanotechnology we design will eliminate both catastrophic and chronic problems, but history suggests that no technology avoids this trade-off.
5. **Control vs. Freedom**: Imagine a terrorist planning an attack that will harm thousands. Fortunately, nanotechnology-enabled drones too small to see have been distributed around most public places. They relay information about the plot to government agents, who capture the terrorists before anyone is harmed. Or, a totalitarian government exploits natural resources for the benefit of the few in power, squashing all dissent through Big Brother-style surveillance of everyone, made possible by the same nanotechnology in the first scenario. Nanotechnology allows us to choose control over freedom. Or, it might let people communicate anonymously and from anywhere, allowing organization for action. Nanotechnology undermines control, giving freedom. When developing a new technology, it may not be clear which effect it will have, or even if it will have any effect on this trade-off of control vs. freedom. But we, as a society, have a better chance of creating the environment we want if we are aware that nanotechnology may allow different trade-offs along this spectrum.
6. **Progress vs. Obsolescence**: Cell phones usually celebrate their second birthday in recycling or trash. The quicker cell phones improve (whatever constitutes "improvement"), the quicker we upgrade and dispose of the old. Cell phones are not modular like computers, with slots for new memory, cards, and storage drives, but even computers quickly become obsolete. Memory standards change (DDR to DDR2 to DDR3), expansion cards change format (AGP to PCI to PCI-e), and storage drives change interfaces (PATA to SATA to SATA-2), making it expensive or impractical to upgrade old systems. So we throw them away, and will do so more rapidly as new models come out more rapidly. Will nanotechnology change this trade-off? Probably not, but it may be able to lessen the impact of waste. Currently, many products

> are not recyclable. Cell phones and computers have many different materials bonded together, and separating them for reuse is expensive in cost, time, or environmental impact (much electronic waste ends up in minimally regulated areas in Southeast Asia, causing pollution even when recycled). If matter compilers could assemble whatever we want from simple molecules, could they also disassemble? Atoms do not wear out, and many molecules stay the same from production through use and disposal. Perhaps nanotechnology will be designed that can disassemble our unwanted products, much as children disassemble Lego models, putting the plastic blocks back in the holding trays for next time. Imagine no material waste. If energy comes from a clean source, there will be little waste at all.

K–12 students are learning how the world works. They may not see how much of what they learn will be useful. They need prompting and time to reflect on the trade-offs of costs and benefits they learn, so they can imagine how they will use it. This reflection leads to values. What do they find important? We can help them reflect on that, too.

How Do We Evaluate Nanotechnology?

Students may be surprised to find that, when analyzing a technology with ICE-9, two of them may agree on the first eight questions, right up through what the costs and benefits are, but disagree about whether the technology is good. Our values tell us whether something is good and those values are subjective. The Greek philosopher Socrates said that the unexamined life is not worth living. We might paraphrase to say that unexamined values are not worth applying. Students will be much better at critical thinking if they examine their own values, even noting how they may differ depending on the situation (e.g., classroom to sporting field to home). Students will also be better at communicating with others because they will be able to recognize other's values. If they accept that values may not be right or wrong, but simply different, they may have less conflict, too. If they are unaware of differing values, then someone else's differing conclusion about whether something is good could appear either malicious or stupid. Neither of those conclusions is conducive to communication, understanding, or cooperation.

Students build on previous activities by taking a familiar tool, the *PMI Chart,* and applying it in a new way. This time, they will fill it out for someone else's perspective. They may do this for a friend or family member. Or the teacher may connect the activity to a reading from another part of the curriculum. Historical readings would allow a student to imagine how someone from the past would evaluate a current technology, or even a future nanotechnology. The ICE-9 curriculum also offers a reading based on a variant of Maslow's Hierarchy of Needs called "Spiral Dynamics." The reading offers categories of values that can be seen in individuals, organizations, and cultures. A very brief synopsis of the reading (a paragraph instead of a page for each value category) is here:[3]

1. **Survival**: *Does it feed and protect us?* Robinson Crusoe gives us a fictional example of a shipwreck survivor on a desert island. When retrieving goods from his foundering ship, he selects pistol, saw, and hammer over gold because food and protection were most important.
2. **Ritual**: *Is it consistent with our myths and traditions?* The Yir Yoront aborigines in Australia evaluated steel axes as good and canoes as bad based on mythology. The steel axes introduced by missionaries were very similar to stone axes, which they believed the gods had given them. But gods had given them nothing like the canoes that their neighbors used.
3. **Power**: *Does it give us more control?* In Afghanistan, the Taliban evaluated Stinger missiles, AK-47 assault rifles, and pickup trucks as good because those maintained control. They evaluated television and satellite dishes as bad because those undermined it.
4. **Authority**: *Do our leaders or traditions say it is good?* China, guided by the authority of Confucian principles, developed mechanical clocks and vast oceangoing fleets of ships long before Europeans did. But a new regime declared new priorities, scrapping much of the advanced technology and turning inward to build the Great Wall.
5. **Economic**: *Is it profitable?* In the capitalist free market system, electric lights, nuclear power, and Internet technologies promised profit and, so, were considered good. Labor-intensive, small-scale agriculture is less efficient than automated, large-scale agriculture, so the market puts the former out of business.
6. **Fairness**: *Does it benefit everyone and everything?* The U.S. government evaluates hydroelectric dams based on how they affect water quality, fish, wildlife and botany, public recreation, historical and archeological sites, and aesthetics.

Students share with the class their selection of people or stories for their PMI Chart. The teacher can list values on the board, looking for commonalities and for those that do not fit in the six categories illustrated in the reading. Reflective journaling is on how evaluating technology might help the student make successful decisions about:

- What they spend their money on
- What they study

- What kind of work they do
- How they lead their lives

We are asking students to prepare for large challenges. Those thinking about nanotechnology hold a wide range of beliefs about how we should consider costs and benefits and how we should consider the values of others. All of our decisions about nanotechnology will involve uncertainty and risk. Risks vary in likelihood, impact, and reversibility. Those most likely, with greatest impact and no chance of reversing in case of poor outcome, deserve our greatest attention. The three dimensions of consequence described in the previous section raises another issue. Those indirect consequences impact people or environments that have little or no say in whether a given nanotechnology would be developed. Humans have much lower tolerance for risk when they do not control the activity. For instance, skiing or bicycling down a steep mountain is risky, but many choose it as recreation. We would be angered and outraged if someone else put us in similar danger without our consent. Liability laws give recourse when such indirect costs are imposed.

The precautionary principle[4] is one policy that minimizes costs imposed on those not directly involved in the decisions to develop or deploy a new nanotechnology. In an essay called *The Precautionary Principle in Nanotechnology*,[5] John Weckert and James Moor propose the approach, "If an action A poses a credible threat P of causing some serious harm E, then apply an appropriate remedy R to reduce the possibility of E." It sounds abstract, but it boils down to trying to anticipate the likelihood and severity of the consequences of an action, and then acting to minimize them. It finds middle ground between "do nothing unless proven safe" and "do anything unless proven dangerous." The first is paralyzing, and the second is destructive.

When technology was less powerful, the consequences of trying anything not proven dangerous were small. The inventor might kill him or herself, and perhaps the people standing too close. Nanotechnology has tremendous power for creation or destruction, so many are concerned, particularly about the indirect, unanticipated costs. Understanding the context so we can assess expert opinion is becoming more important, because those suffering a negative outcome will no longer be only those standing near the developer, but possibly on the other side of the world and generations later.

In K–12, we will not solve the challenges posed by nanotechnology, but we can equip our students for critical thinking so that they can take on aspects of these challenges. We can help students understand what nanotechnology is, why we use it, where it comes from, and how it works. We can show them how it changes, how it changes us, and how they can change it. We can share patterns in costs and benefits and help them reflect on their own values.

Society tries to restrict powerful technology from those not yet mature enough to wield it. But the rules for when you can drive a car or buy a gun are largely measures of age, not maturity, wisdom, insight, or ability to think critically. Control at the nanoscale brings technology far more powerful than cars or guns. K–12 education is perhaps the only place that we can prepare those who will not only develop tomorrow's nanotechnology, but also question it, pass laws about it, invest in it, manage it, and promote it. In the future, nanotechnology may affect everyone, so K–12 needs to prepare all students with some level of literacy about nanotechnology, and the ability to think critically about it. Some students will build on this foundation to study the advanced science, math, and engineering to create that nanotechnology, but we should all be capable of understanding and evaluating it.

References*

1. AAAS Project 2061, Chapter 3: The Nature of Technology, http://www.project2061.org/publications/sfaa/online/chap3.htm
2. Miguel F. Aznar, *Technology Challenged,* Knowledge Context, Santa Cruz, 2005, p. 126.
3. ICE-9 curriculum download, Teacher Guide, p. 54, http://knowledgecontext.org/Curriculum/download.htm
4. Precautionary principle, http://en.wikipedia.org/wiki/Precautionary_principle
5. Fritz Allhof, *Nanoethics: The Ethical and Social Implications of Nanotechnology,* Wiley Interscience, Hoboken, New Jersey, 2007, p. 144.

* All links active as of August 2010.

Section III

Nanoscience Resources and Programs

Science may set limits to knowledge, but should not set limits to imagination.

Bertrand Russell (1872–1970)

This section of the book provides short overviews and links to outreach resources for teachers/students in grades K–12, and teacher training programs. Some of the centers are part of larger networks that provide informal learning, reviewed in Chapter 11, featuring museum exhibits, lectures, cable TV stations, and Internet Web sites. Chapter 10 is dedicated to summaries of workforce programs developed in five states that can serve as models for replication for training of nano technicians. Chapter 12 provides a synopsis of developing nations' nanotechnology initiatives, identifying niche training programs in the global arena and educational resources available for global access.

9

K–12 Outreach Programs

A five-year goal of the NNI is to ensure that 50% of U.S. research institutions' faculty and students have access to the full range of nanoscale research facilities, and student access to education in nanoscale science and engineering is enabled in at least 25% of the research universities.

Mihail C. Roco
NSF Senior Advisor for Nanotechnology[1]

Overviews: Nanoscience Education Outreach Programs from U.S. Universities and Nano Centers

A highly significant impact of the National Nanotechnology Initiative (NNI) is the focused investment by the eight participating agencies[2] in the establishment and development of 85 multidisciplinary research and education centers devoted to nanoscience and nanotechnology. Many centers, with state-of-the-art equipment for nanoscale research, are designated user facilities, available to researchers from academia, the private sector, and scientists at national laboratories.

The resources provided by these Nano Centers and Universities are designed as outreach to inform the public along with teachers and students. The goal was to provide a complete resource overview as the beginning of a Clearinghouse for K–12 Nanoscience Education. Please accept our apologies for any program that was not included.

The Institute for Chemical Education, Madison, Wisconsin

Under the leadership of Professor John W. Moore, W. T. Lippincott Professor of Chemistry, this center has been an early advocate for e-learning to advance science education. Understanding the power of interactive education, Professor Moore was already working on the National Science Digital Library project when he and I were introduced in 1998.

In addition to his work with the Institute of Chemical Education (ICE), Professor Moore has made major contributions to several important projects in chemical education along with coauthorship of textbooks: The *Journal of Chemical Education,* as founder in 1989, and editor (1996–2009); the National Science Digital Library—he heads one of 11 Pathways projects, Project SERAPHIM, to develop innovative chemical education materials using computers and other technology.

Institute for Chemical Education: Overview

The Institute for Chemical Education (ICE) was established in 1983 to provide a center for science educators to develop and disseminate ideas. ICE is national in scope and has led the drive to help teachers revitalize science in schools throughout the United States. Its efforts include creating new materials for teaching and learning, research in chemical education, development of demonstrations and hands-on activities, sponsors of workshops for teachers, organization of laboratory and demonstration programs for school children, and dissemination of exemplary ideas via publications and kits.

All ICE programs emphasize hands-on science, taught interactively to help students develop powers of observation and problem solving. ICE aims to stimulate the scientific curiosity of all teachers and students, not just those traditionally well served by our educational system, through these goals:

- Make science hands-on at all levels
- Enable teachers to become better teachers, not merely tell them what to do
- Draw ideas from what teachers say they need
- Develop and distribute materials that are usable and affordable for the average teacher in the average classroom

Over 3500 teachers of kindergarten through college have attended workshops at ICE headquarters, located at the University of Wisconsin–Madison, and at other locations across the country. In addition, ICE affiliates at more than 60 other colleges and universities have been trained to carry out workshops that enhance the teaching of science.

ICE also serves as the Education and Outreach arm for the University of Wisconsin–Madison Nanoscale Science and Engineering Center on Templated Synthesis at the Nanoscale (NSEC). The NSEC is one of 16 centers funded through the National Nanotechnology Initiative to build infrastructure for nanotechnology research and education across the country.

The NSEC Education and Outreach program develops new scalable teaching and learning programs, methodologies, and communities to cultivate a diverse next generation of nano scientists and engineers. They educate underrepresented groups, offer teacher professional development for K–12 schools, and design new teaching models for students with physical and learning disabilities.

ICE programs[3] receiving support from the NSEC include: SCI ENCountErs for K–12, informal science education exhibits, teacher workshops, independent laboratory access for the blind, Today's Science for Tomorrow's Scientists program, online nanoscience for teachers, and research experience for undergraduates.

Support for ICE programs comes from the University of Wisconsin–Madison and through grants from National Science Foundation.

Exploring the Nanoworld

The University of Wisconsin–Madison Materials Research Science and Engineering Center (UW MRSEC) Interdisciplinary Education Group, under the direction of Wendy Crone, works in unison with ICE. Using examples of nanotechnology and advanced materials to explore science and engineering concepts at the college level, they manage to bring the "wow" and potential of nanotechnology and advanced materials to the public. The curriculum developed under their program for K–12 students follows the state and national standards of excellence, and passes all assessment testing requirements. The programs contain lesson plans, units and courses, background and supporting slides, videos of laboratory experiments, laboratories, activities, demo videos, exhibits, and kits.

Grades 6–12

The *Nanoworld Cineplex*[4] has been one of the most popular online resources for teachers for over a decade as a demonstration tool in their classroom. The *Nanoworld Cineplex* facilitates instruction in science and technology using a visual medium to show basic experiments related to nanotechnology.

The Guide to the Curriculum Materials[5] page has categorized listings for all the resources as a menu that teachers and students can explore.

The most recent development under the Kits section is a board game, developed for the students and parents to explore connections between science, nanotechnology, and society.

NanoVenture: The Nanotechnology Board Game[6]—High School

NanoVenture: The Nanotechnology Board Game explores the connections between nanoscience, nanotechnology, and society. In this game, players become leaders of a new country. The leaders are challenged to make decisions regarding their country's use of nanomaterials and nanotechnology for industrial expansion, military applications, economic security, and basic scientific research, while maintaining a high approval rating from the citizens of the country. These decisions require players to carefully analyze the interplay of technological advances, regulations, public perception, and risk, while also learning about the emerging field of nanotechnology.

Materials World Modules at Northwestern University—Middle and High School

The NSF-funded *Materials World Modules (MWM) Program*[7] has produced a series of interdisciplinary modules based on topics in materials science, including composites, ceramics, concrete, biosensors, biodegradable materials, smart sensors, polymers, food packaging, and sports materials.

Two new modules for teaching nanoscience and nanotechnology are now available for teachers.

The modules are designed for use in middle and high school science, technology, engineering, and math classes; and are used by over 35,000 students in schools nationwide.

The MWM modules are based on principles of inquiry and emphasize active, hands-on learning. Most importantly, they provide middle and high school students new opportunities to apply what they learn in the classroom to real-world problems. MWM modules are designed to meet all National Science Education Standards and state standards.

MWM strives to offer user support through informational workshops, training programs, networking, and mentor support. They also offer exhibits of classroom projects, a listing of helpful online resources, and a Web-based collaboration network that includes an online question and answer service for educators and students. The network *MWM NET*[8] is a virtual community for teachers across the nation and around the globe to interact and share resources through the MWM Web site. Members can network with fellow MWM teachers and view and contribute to the extensive Web-based gallery of free teaching resources and services.

NCLT—National Center for Learning and Teaching Nanoscale Science and Engineering—Northwestern University

The NSF-supported National Center for Learning and Teaching in Nanoscale Science and Engineering (NCLT)[9] is designed to build capacity in Nanoscale Science and Engineering Education (NSEE). They aim to equip future generations by advancing science, technology, engineering, and mathematics (STEM) education. Established in October 2004, the guiding theme of the

NCLT is research on learning and teaching through inquiry and design of nanoscale materials and applications. On their *NanoEd Resource Portal*[10] an explanation of their progress since 2004 to identify appropriate introduction of nanoscience concepts into science curriculum has produced four reports that you can download in PDF files on the Web site:

- *Exploration of Student Understanding and Motivation in Nanoscience* (PDF)
  Kelly Hutchinson, Namsoo Shin, Shawn Y. Stevens, Molly Yunker, César Delgado, Nicholas Giordano, and George Bodner
- *The Development of Students' Conception of Size* (PDF)
  César Delgado, Shawn Y. Stevens, Namsoo Shin, Molly Yunker, and Joseph Krajcik
- *Using Learning Progressions to Inform Curriculum, Instruction and Assessment Design* (PDF)
  Namsoo Shin, Shawn Y. Stevens, César Delgado, Joseph Krajcik, and James Pellegrino
- *A Design-Based Approach to the Professional Development of Teachers in Nanoscale Science* (PDF)
  Lynn Bryan, Shanna Daly, Kelly Hutchinson, David Sederberg, Randal Batchelor,[2] Eric Hagedorn, William Fornes, and Nicholas Giordano

Lesson Plans[11] for Teachers

Level: Grade 7–12 Teachers

Unconventional Patterning at the Nanoscale
Submitted on 23-FEB-07
Contributed by Teri Odom, Meenakshi Viswanathan, Yelizaveta Babayan

The ability to generate nanoscale structures is central to modern science and technology. The experiments described in this site provide the first step toward bringing nanofabrication to undergraduate students using simple benchtop tools that are accessible and inexpensive. These experiments are currently taking place in the first quarter of a sophomore seminar (Chem 250-1) offered jointly by the Weinberg College of Arts and Sciences and the McCormick School of Engineering at Northwestern University.

What Can Electrons Do?—Electron Microscopy
Submitted on 13-FEB-07
Contributed by Jian-Guo Zheng

This project is to educate students about important roles of electron microscopy in nanoscience and nanotechnology. It will cover some properties of electrons, principles of electron microscopy, applications of electron microscopy in characterizing nano-scale materials, and the development of electron microscopy to meet challenges in nanotechnology.

Nanomaterials, MSE 376 (Spring 2005)
Submitted on 02-MAY-06
Contributed by Mark Hersam

Materials Science and Engineering 376, Nanomaterials, is an interdisciplinary introduction to processing, structure, and properties of materials at the nanometer length scale. The course will cover recent breakthroughs and assess the impact of this burgeoning field. Specific nanofabrication topics include epitaxy, beam lithography, self-assembly, bio-catalytic synthesis, atom optics, and scanning probe lithography. The unique size dependent properties (mechanical, thermal, chemical, optical, electronic, and magnetic) that result from nanoscale structure will be explored in the context of technological applications including computation, magnetic storage, sensors, and actuators.

Nanocos: The Card Game of Nanotechnology Concepts[12]

Level: Middle School and Above

Brief Summary of the Game

Nanocos is a two-player card game, with each player having a deck consisting of Object cards, Action cards, Microscope cards, and Carbon cards. Taking turns, each player selects an Object card to engage with the opponent's Object card; the player with the larger SA/V prevails, and he or she gains a Carbon card. The complexity of the game is multiplied by the use of Action and Microscope cards. Some Object cards require the use of Microscope cards in order to be seen. The first player to collect all five Carbon cards, representing the five allotropes of carbon, wins the game.

Learning Goals of the Game

To relate volume and surface area for different objects that students have already learned in chemistry, biology, and astronomy and extend this concept to the nanoscale:

- To understand size scales
- To make calculations with powers of ten
- To realize that smaller objects have larger surface area to volume ratios, which can result in drastic physical or chemical property changes at the nanoscale

- To know different types of microscopes and what size objects they are designed to view
- To use the scientific method (hypothesizing, observing, drawing conclusions)
- To employ strategic play

The NCLT is part of the group of NSF-funded Centers for Learning and Teaching;[13] CLTnet[14] lists resources and reviews of all centers.

The NCLT in Nanoscale Science and Engineering Has Partnered with Taft High School of the Los Angeles Unified School District to Launch a Nanotechnology Academy

Funding from the U.S. Dept. of Education "Small Learning Community" initiative and an SSP Nanotechnology grant from the State of California supports the program. The Academy will train several hundred of the 3,000 Taft students over a four-year period using an integrated curriculum that links nanotechnology to a wide range of courses including science, technology, engineering, math, social studies, language arts, and fine arts. NCLT is partnering with 146 Taft teachers and administrators to develop and implement the new curriculum, offer summer research opportunities, and provide professional development for the teachers already certified in math and science.

NCLT's new UCLA Regional Hubsite combines the resources of California Nanosystems Institute (CNSI), UC Center for Environment Implication of Nanotechnology (CEIN), and the new $25 million NSF/EPA–funded center to provide year-round mentoring for the project, with input from local industrial researchers.

The one-year *Passport Course* features five units of inquiry-and-design-based NCLT nano-modules and classroom kits, enhanced with Web-based animations, simulations, and games. A virtual lab space enables Taft students and teachers to collaborate with each other and with their peers at other NCLT-network schools to be developed nationwide.

Three more NCLT Hubsites have been established: Central U.S.: Colorado State University–Center for Extreme Ultra Violet (EUV); Southern U.S.: Louisiana State University–Gordon A. Cain Center; and Northeastern U.S.: State University of New York–Albany. The original NCLT at Northwestern University will remain the Hubsite for the Midwest, with two more locations to be named in the future. The Network of NCLT Regional Hubsites will receive infrastructure support, capacity building, and joint funding. Hubs will provide expertise, field test data, professional development, and classroom outreach. Hubs will connect colleges, school districts, NSE Centers, national labs, industry, and museums.

Ohio State University–Center for Affordable Nanoengineering of Polymeric Biomedical Devices (CANPBD)

The CANPBD goal is development of polymer-based, low-cost nano engineering technology, to produce micro-/nanofluidic devices, cell-based devices, and multifunctional polymer-nanoparticle-biomolecule nanostructures. These devices are needed for production of next-generation medical diagnostic and therapeutic applications.

Outreach Educational Programs for Students

Throughout the school year, groups of students from area high schools and home-schooled students visit the labs for tours and hands-on experience with state-of-the-art equipment.

Teacher's Workshops

Summer: A full-day, hands-on workshop for high school teachers that includes a take-home kit with everything required to do each experiment in the classroom. Sixteen teachers from high schools in and around central Ohio participated in a 2009 workshop, learned about nanotechnology, and observed and performed experiments. Experiments ranged in complexity from measuring a nanometer to modeling tissue engineering using nanotechnology. The teachers also toured the Nanoscale Science and Engineering Center clean room facility at Nanotech West and observed researchers at work. Afternoon activities included a cryogenics and phase change demonstration: the preparation of liquid-nitrogen ice cream.

Teacher Resources[15]

Currently 16 lessons: high school teachers can download these free online.

The College of Nanoscale Science and Engineering (CNSE) of the University at Albany[16]

This is the world's first college dedicated to research, development, education, and deployment in the emerging disciplines of nanoscience, nanoengineering, nanobioscience, and nanoeconomics. With more than $5.5 billion in public and private investments, CNSE's Albany NanoTech Complex has attracted over 250 global corporate partners.

Students, faculty, and corporate partners use leading edge tools at CNSE's Albany NanoTech Complex, which houses research and development, and prototype manufacturing infrastructure, Class 1-capable clean rooms, and the most advanced 200-mm/300-mm wafer facilities in the academic world.

CNSE students take part in a groundbreaking course of study. The NanoCollege's unique, cross-disciplinary curriculum, offering B.S., M.S., M.S.–M.B.A. and Ph.D. degree programs, allows students to delve deeply into a wide variety of advanced topics needed to support cutting-edge research.

CNSE's Albany NanoTech Complex serves as an acceleration facility for more than 250 global corporate partners, helping them overcome technical, market, and business development barriers to successfully deploy nanotechnology-based products.

High-school seniors[17] conducting research at CNSE receive recognition in the prestigious Intel Science Talent Search. From clean energy initiatives to revolutionary health-related research, three high school students took their science projects to the next level under the guidance of CNSE faculty as part of internships at CNSE's world-class Albany NanoTech Complex.

New Bachelor's Degree Program

The bachelor's degree in Nanoscale Engineering at CNSE complements the undergraduate degree in Nanoscale Science—launched last June as the world's first comprehensive baccalaureate degree program in nanoscale science—as well as CNSE's graduate-level curriculum, which was also a global first when it commenced in September 2004.

Students completing the CNSE bachelor's degree program in Nanoscale Engineering will be well-equipped to solve challenging problems in nanoscale engineering and applied sciences, as well as other physical and biological engineering sciences, and will be uniquely educated to pursue opportunities in emerging high-technology industries, including nanoelectronics, nanomedicine, health sciences, and sustainable energy, as well as competitive graduate degrees in emerging engineering fields.

The Nanoscale Engineering curriculum comprises a cutting-edge, inherently interdisciplinary academic program centered on scholarly excellence, educational quality, and technical and pedagogical innovation. The program will give students the abilities to creatively solve nanoscale engineering problems through the use of rigorous analytical, computational, modeling, and experimental tools based on foundational physical or biological sciences and mathematics, nanoscale materials, and applied engineering concepts.

The program will also enable students to propose, formulate, and execute original, cutting-edge nanoscale engineering research at CNSE's Albany NanoTech Complex. UAlbany NanoCollege offers students a world-class experience working with, and learning from, the top innovative minds in the industrial and academic worlds.

The CNSE houses the only fully integrated, 300-mm wafer computer chip pilot prototyping and demonstration line within 80,000 square feet of Class 1-capable clean rooms. More than 2,500 scientists, researchers, engineers, students, and faculty work on site at CNSE's Albany NanoTech, from companies including IBM, AMD, Global Foundries, SEMATECH, Toshiba, Applied Materials, Tokyo Electron, ASML, Novellus Systems, Vistec Lithography, and Atotech.

Nano for Kids[18] Programs

CNSE conducts multilevel programs that encourage science awareness in grades K–12 to help ensure the U.S. science and technology workforce of the future.

NanoCareer Day

NanoCareer Day introduces middle and high school students to the exciting world of nanotechnology through tours, presentations, and hands-on activities.

NanoHigh

NanoHigh, developed jointly by the City School District of Albany (CSDA) and CNSE, offers students an unprecedented opportunity to study the emerging field of nanotechnology through two courses that combine classroom learning at Albany High School with monthly on-site visits to CNSE for in-depth lab activities.

Excelsior Scholars Nanoscale Science Summer Institute—Grade 7

Funded through the New York State Education Department's Excelsior Scholars Program, the Nanoscale Science Summer Institute provides a one-of-a-kind, hands-on opportunity for seventh grade students to take part in advanced academic coursework that incorporates cutting-edge nanoscale science research.

NanoEducation Summit

CNSE recently held the region's first-ever Nano Education Summit for school superintendents, administrators, and teachers, with discussion focusing on ways to implement nanotechnology awareness and learning at the middle and high school level.

Regional Collaboration

CNSE partners with a variety of regional organizations to promote nanotechnology awareness among teachers, students, and parents. Some recent partnerships include:

- Capital District Future City Competition
- Tech Valley Summer Camp
- Girl/Boy Scouts
- Alliance of Technology and Women Great Minds Program

School Presentations and Tours

CNSE hosts a variety of school groups throughout the year—from administrators and faculty—to students and their parents. The faculty, staff, and graduate students at CNSE take great pride in providing access to state-of-the-art facilities and a world-class team in an effort to educate the youngest members of our community. CNSE's educational outreach programming inspires students to pursue careers in science and technology, creating the innovative workforce that is critical to our future on a regional, statewide, and national scale.

Nano Games[19] Grades 6–12

NanoMission: Action Adventures in the Nano World

Try your hand at one of these exciting "nano games," courtesy of *PlayGen's NanoMission*™—the world's first scientifically accurate interactive 3D learning game based on understanding nanosciences and nanotechnology!

To play the games, register and download the modules.

Columbia University–MRSEC Center for Nanostructured Materials, New York City (NYC)

NYC High School Visitation Program

Programs were developed to bring MRSEC faculty and students to NYC schools to promote materials and technology with laboratory demonstrations in materials science, explained through exciting, hands-on demonstrations incorporating "everyday" objects.

Goals: Fostering the interests of high school students in science and technology by showing them the marvels of materials science and engineering, describing the relevance of science to quality of life, and improving the retention of a diverse student body to stay on track for careers in science and technology.

Target audience: Developed for grades 8–12

Frequency offered: Approximately four schools are visited per semester, with a total of eight schools visited per calendar year.

Ron McNair Curriculum Integration To Interactively Engage Students (CITIES) Program

Started in 2003, the program expanded from solely working with teachers, to creating engaging lab demonstrations, to helping teachers and students with their school's internal science fairs.

Research and Rolling Exhibits (RARE)

In 2005, the Columbia MRSEC and the New York Hall of Science received a grant from the National Science Foundation to develop five "Discovery Carts."

Requirements: Designed as visually attractive portable units, intellectually stimulating, with demonstrations of current research in materials science and nanotechnology, and available for daily public use on the main floor of the museum.

Research Experiences for Teachers (RET)[20]

The Research Experiences for Teachers (RET) program enabled four grade 9–12 teachers to do research in the MRSEC each summer since 1999. The teachers help evaluate the MRSEC education and outreach programs to NYC K–12 schools.

Frequency offered: 10 weeks every summer.

Goal: Provide New York City public high school science teachers with hands-on experience in scientific research, expanding their knowledge base and proficiency to teach students and fellow teachers. Each MRSEC RET participant spends two consecutive summers working as a laboratory research assistant under the supervision of a Columbia MRSEC faculty mentor. They participate in weekly Monday meetings with other teachers and science and education professionals. Program is designed to encourage professional development, scientific understanding, and communication, science teaching, peer coaching and technology development. At the end of the summer symposium, the first-year teachers give a poster presentation and second-year teachers present an oral presentation.

University of Pennsylvania MSREC

Lectures for Science Teachers Series[21]

Advanced Materials: Synthesis, Characterization and Properties: Composed of a monthly lecture, by a faculty member affiliated with the Laboratory for Research on the Structure of Matter (LRSM). Each lecture is followed by a visit to one of the Shared Experimental Facilities of the LRSM considered pertinent to the lecture topic. Approximately 20 teachers from the Philadelphia School District attend this noncredit course annually at Laboratory for Research on the Structure of Matter.

All lectures are free, take place on a Thursday once a month at 5:30 p.m., and are followed by refreshments.

Penn Summer Science Initiative

The Penn Summer Science Initiative (PSSI) program for 2009 enrolled 29 high school students from the Delaware Valley. The class consisted of 80% public and 20% private and parochial school students; 17% of the students were minorities and 62% were females.

A series of 15 lectures covering the major classes of materials were given by seven faculty members and two guest lecturers from DuPont. A series of four experimental labs, each requiring three afternoon sessions, were supervised by graduate students, supported by the Penn MRSEC and five staff members of the LRSM. Students were required to write lab reports for evaluation and finished the program with joint oral presentations to receive certificates of participation. The program included field trips to relevant facilities and museums.

This program was restricted to commuters and was free to all students with emphasis on attracting local public school students. Strengths and weaknesses of the program were evaluated, and adjustments will be made for the program in 2010.

Lehigh University–Outreach K–12

ImagiNations[22]

ImagiNations is an outreach program of the Center of Advanced Materials and Nanotechnology (CAMN), Lehigh University. The program features an interactive microscope with "Introduction to Nanotechnology" for students and teachers. The Internet site features a section that explains electron microscopy and the first nanotechnology textbook for grades 1–6,

Nanotechnology for Grades 1–6+, Introducing Nan and Bucky Dog, by Andrea J. Harmer. K–12 teachers can bring students on an exciting field trip to the nanotechnology research labs, or run the XL30 electron microscope from their classrooms.

Arizona State University's Interactive NanoVisualization for Science and Engineering Education (IN-VSEE)[23] Project Initiated in 1997

IN-VSEE is a consortium of university and industry scientists and engineers, community college and high school science faculty, and museum educators with a common vision of creating an interactive Internet site for education based on remote operation of advanced microscopes and nano-fabrication tools, coupled to powerful surface characterization methods.

IN-VSEE was funded by the National Science Foundation from 1997 to 2002. This was followed by a period of time (2002–2007) where IN-VSEE resources continued to be used by elementary, middle school, high school, and college educators. During this period there was minimal support for the program, but thanks to recent successful grants, IN-VSEE has a renewed emphasis. New efforts are being put toward modernizing the Web site with new, more modern material, updating Java and Internet browser compatibilities, and increasing reliability.

Preparing students for the workforce in the imminent nanotechnology revolution is a goal of the online interactive virtual lab. The IN-VSEE initiative has been a trailblazer in the integration of research, education, and outreach utilizing a virtual laboratory on the Internet. The highlight of the experience is the SPM—LIVE! Designed as the heart of the virtual lab, it gives educators and researchers the opportunity to perform scanning probe microscopy (SPM) experiments, with image broadcasting and control of the instrument on a real-time basis. This allows access to state-of-the-art instrumentation that may not be available at their institution during a lecture or laboratory session. Groups that are authorized to control the experiments can use samples available at IN-VSEE or submit their own after review.

Modules for Teachers[24]

The Modules are developed for teachers/students to understand microscopy before they use the SPM-LIVE! Section. In the Module section, brief summaries are provided for teachers:

Making Sense of Scale and Size—Key Concepts and Learning Objectives

> It is essential that a student of science and/or engineering understands the intimate relationship between scale and size, micro/nanoscopic structure and the properties of materials and their function at the macroscopic level. Take a look at the microscopic world through optical, scanning electron and scanning probe microscopes in this module, and learn some astonishing facts about how scale and size affect properties and hence functions of substances in our material world.

A central theme that ties the modules together is that the structure, properties, processing, and performance parameters of a material are intimately linked at all levels of scale.

The interactive IN-VSEE modules provide:

- Reinforcement of key concepts and fundamental principles that are taught in science, math, and engineering curricula.
- Prospective user lessons on the methodologies of experimental design and remote SPM instrumentation on the Internet.
- The modules will challenge and encourage potential users to formulate original experiments.

The educational modules were written in collaboration with teachers from precollegiate and collegiate institutions. They center around a common theme of understanding and manipulation of our natural and man-made material worlds. The module topics selected rely heavily on interactive, discovery-based learning activities to introduce or reinforce the user. They learn the key fundamental and applied material concepts within a material class or discipline, plus various material classes and disciplines over a wide range in length scales.

Image Gallery

This gallery is a unique bank of images developed under the aegis of the IN-VSEE project. These images illustrate the interdisciplinary nature of modern science and engineering and the value of integrating research into education. They represent a cross-section of disciplines and across ranges of size and scale. They are in the forms of photographs, schematics, animations, and micrographs taken from scanning electron microscopy (SEM), scanning probe microscopy (SPM), optical microscopy, and surface analytical techniques. Contact the university for ongoing teacher workshop schedules.

Down to Earth GK–12[25]

Down to Earth Science is a collaborative project under the direction of B. L. Ramakrishna, Ph.D., that brings together Arizona State University scientists, engineers, and graduate students to enrich learning experiences for the K–12 community. The main goals of this project are to improve communication and teaching-related skills for graduate fellows, strengthen partnerships between ASU and the K–12 community, and provide new opportunities for K–12 students and teachers to work with practicing scientists and engineers.

18 lessons posted under Physical Sciences[26]—size and scale

43 lessons posted under Life Sciences[27]

Lessons plans[28] for Anthropology, Earth and Space Sciences, Science and Society, General Science

Georgia Institute of Technology–Nanotechnology Research Center

Expertise: Biology, Life Sciences, Integrated Systems, Electronics

The Georgia Tech-NNIN (GIT-NNIN) site emphasizes the application of nanofabrication to bioengineering and biomedicine. Much of the GIT-NNIN activity is carried out in the Nanotechnology Research Center (NRC), housed in the Marcus Nanotechnology Building. The major fabrication facilities include thin film deposition, plasma processing, optical and electron beam lithography (100 keV/4-nm spot size JEOL JBX 9300FS), thermal processing, electroplating, wafer lapping, polishing, dicing, bonding and sawing, wire bonding, flip chip bonding, III-V MBE growth, a 2-metal/2-poly CMOS line, and a MEMS line (ion implantation outsourced).

Characterization tools include optical and electron microscopy, X-ray tomography, mass-spectroscopy, AFM, SCM, STM, stylus and optical profileometry, low/high force scanning nanoindentation tribology, surface analysis tools, and high-speed electronic and optical testing. Design and simulation tools are available on PC and workstation clusters under campus-wide site licenses. Other GIT-NNIN resources include the GIT Electron Microscopy Center and the Laser Dynamics Lab (LDL). These facilities are open to a NNIN user community from a wide range of disciplines including electrical, computer, mechanical, chemical, materials, and biomedical engineering, as well as physics, chemistry, and biology.

The Microelectronics Research Center (MiRC) at Georgia Tech provides expertise, facilities, infrastructure, and teaming environments to enable and facilitate interdisciplinary research in microelectronics, integrated optoelectronics, and microsensors and actuators.

MiRC is the location for the Education and Outreach Office of the National Nanotechnology Infrastructure Network (NNIN). NNIN provides a wide variety of educational outreach in nanotechnology that reaches school-aged children through adult professionals. The Georgia Tech NNIN Education and Outreach Office has as its mission to provide educational programs for young people, teachers, undergraduates, graduate students, adult professionals, and the general public. The goals are to ensure a nano-literate public in Georgia and to encourage students to pursue education and careers in science, technology, engineering, or mathematics, and particularly in nanotechnology. The outreach efforts would not be possible without the numerous dedicated volunteers from Georgia Tech and Emory University who assist with the programs.

Instructional Units[29]

20 unit lessons are posted for K–12.

Sample: Size and Scale Unit

An instructional unit designed to help students better visualize the scale of nanoscience. Students learn how small a nanometer is compared to everyday objects.

Unit includes: Teacher Guide, Size and Scale Pictures, Size Units for a Number Line

Nano Camps[30]

Each summer MiRC hosts a week-long summer camp for high school students interested in nanotechnology (nonresidential). The course is designed for rising 10th, 11th, and 12th grade students who have completed high school physical science or chemistry. The camp is held in either late June or early July each summer and is titled "Advanced Topics Course: Nanotechnology Explorations." The program is a combination of hands-on activities, laboratory tours, and lectures. Participants tour research laboratories, interact with world-renowned scientists, and complete hands-on activities designed to introduce them to nanotechnology research and development issues. In the process, students explore chemistry at the nanoscale and examine how scientific tools, including the atomic force microscopes, are used in nanotechnology research.

The MiRC recently expanded its summer *Nanotechnology Explorations* program to satellite locations across the Heart of Georgia RESA Area. These week-long camps are run by Georgia teachers, funded by the Improving Teacher Quality Grant and expand the outreach to more students than ever before.

MiRC also supports the Women in Engineering Program's "Technology, Engineering, and Computing Camp." This camp offers middle school girls an early introduction to the world of technology, engineering, and computing. MiRC provides hands-on experiences about nanotechnology to camp participants.

Nanooze Exhibit[31]

Throughout Fall 2010, the Marcus Nanotechnology Building is home to the Nanooze Exhibit. This interactive exhibition is designed to introduce nanotechnology concepts and help educate visitors interested in learning more about nanotechnology. Six interactive exhibits are on display.

Research Experience for Teachers–Nanotechnology Research Center[32]

This is an eight week program: six weeks at MiRC, one week in-school support, one week at NSTA.

What is all the buzz about nanotechnology? Nanotechnology is believed to be the next great technological revolution which will affect many aspects of our lives from electronics to medicine and from materials to privacy. Come experience this exciting area of research.

The NNIN RET Program is funded by the National Science Foundation under the Research Experiences for Teachers Program. The NNIN program is at five NNIN sites: Georgia Institute of Technology, Harvard University, Howard University, Pennsylvania State University, and the University of California, Santa Barbara.

REU Program[33]

The NRC participates each summer in the NNIN Research Experience for Undergraduates (REU) Program by providing high-quality research experiences for undergraduates from across the nation. REU participants spend 10 weeks working on mentored research projects which utilize our state-of-the-art nanotechnology facility and equipment. Participants are matched with faculty and become active members of a research group. Typically, participants work with one or two graduate students who are part of the research group.

An objective of the NNIN REU program is to provide students access to advanced equipment and training and the opportunity to participate in a state-of-the-art hands-on research project in nanotechnology. The program seeks to promote student excitement toward graduate-level research and encourage students to pursue advanced education and careers in nanoscale science and engineering. Participants are provided support to enhance their communication and presentation skills for the annual NNIN REU Convocation. All participants gather for a national research convocation on nanotechnology. This NNIN REU Convocation occurs at one of the NNIN sites in early August. Each student presents the results of his/her summer work both in oral and poster format. In addition, each participant submits a technical report which become the NNIN REU Research Accomplishments for that year.

NANOFANS FORUM[34]—Focusing on Advanced Nanobio Systems

At Microelectronics Research Center, GaTech, a new Forum called NANOFANS (Focusing on Advanced Nanobio Systems) is being instituted. This will be a biannual forum.

The goal for the forum is to connect the medical/life sciences/biology and nanotechnology communities by reaching out to researchers in the biomedical/life sciences areas, to let them know what nanotechnology can offer them in the advancement of their research.

Nano@Tech[35]—Sharing Our Knowledge, Shaping the Future

Nano@Tech is an organization comprised of professors, graduate students, and undergraduate students from the Georgia Tech and Emory campuses and professionals from the corresponding scientific community that are interested in nanotechnology. Meetings are held on the second and fourth Tuesday of each month during the academic year at noon in the Marcus Nanotechnology Building conference rooms (rooms 1116–1118).

Nano@Tech members also support the National Nanotechnology Infrastructure Network's Education and Outreach Office at Georgia Tech's Microelectronics Research Center.

The K–16 programs are a success due to the growing commitment of a dedicated group of cross-disciplined Georgia Tech engineers and researchers who are members of Nano@Tech. This networking group meets for seminars and generously volunteers their knowledge, research, labs, and clean rooms to help motivate young people about possible education and career opportunities and excite them about nanoscale science and engineering.

Purdue University–Nano-HUB

The nanoHUB is a Web-based initiative spearheaded by the NSF-funded Network for Computational Nanotechnology (NCN). The NCN has a vision to pioneer the development of nanotechnology from science to manufacturing through innovative theory, exploratory simulation, and novel cyber-infrastructure. The Network for Computational Nanotechnology is a network of universities that work together to define, develop, and support the nanoHUB. Collaborators and partners across the world have joined the NCN in this effort.

The site is free, with access to online simulation tools enabling individuals to develop learning modules or create lessons. A search of the site with K–12[36] key words found 40 lessons. They are not categorized by grade-appropriate level. Teachers would find the site structure very time consuming with a need to review each lesson for subject, content, and separate teacher materials which were not available. Pedagogy information and inclusion of National and State Standards of posted lessons would also need to be clearly defined for teachers to use the site.

Harvard University–Nanoscale Science and Engineering Center (NSEC)

A collaboration between Harvard University, the Massachusetts Institute of Technology, the University of California, Santa Barbara, and the Museum of Science–Boston, with participation by Delft University of Technology (Netherlands), the University of Basel (Switzerland), the University of Tokyo (Japan), and the Brookhaven, Oak Ridge, and Sandia National Laboratories.

The NSEC combines "top-down" and "bottom-up" approaches to construct novel electronic and magnetic devices with nanoscale sizes and understand their behavior including quantum phenomena. Through a close integration of research, education, and public outreach, the Center encourages and promotes the training of a diverse group of people to be leaders in this new interdisciplinary field. NSEC at Harvard is supported by the National Science Foundation.

Project TEACH

NSEC faculty share their enthusiasm for science through Project TEACH (The Educational Activities of Cambridge and Harvard). This early college awareness program is a joint effort of the MRSEC, the NSEC based at Harvard, and the Harvard Office of Community Affairs. Coordinated with the Cambridge Public Schools, Project TEACH brings each 7th grade class (approximately 500 students) from the Cambridge Public School District to Harvard University throughout the school year. During the visit, students receive information about college admissions and learn about college life from Harvard undergraduates. The class visit culminates in an interactive science presentation by an NSEC faculty member on his or her research and its societal benefits. The relationship with the Cambridge Public Schools extends this program outreach at the high school level through collaboration with the K–12 program.

RET Program

The NSEC, in collaboration with an REU/RET Site in Materials Research and Engineering, hosts teachers to work side-by-side with faculty, postdoctoral researchers, graduate students, and REU participants on research or science curriculum projects. Teachers commit to 6 to 8 weeks during the summer and are invited for a second summer to refine educational modules that are developed as a result of their research experience.

RET participants also attend weekly seminars on research topics and on research ethics. The integrated nature of RET and REU activities, particularly the faculty seminars during the summer, provide ample opportunity for teachers to explore development of small classroom modules based on seminar content. Howard Stone's seminar on "Unintended Consequences of Research" was particularly amenable to conversion to the high school classroom, because it presented the progression of science as a nonlinear process that includes many reversals and rewritings, in contrast to the often cut-and-dried presentation in many textbooks. As a result, several teachers adapted this material for their classrooms.

GK–12 Program

The K–12 Program sponsored by the National Science Foundation (NSF) puts Harvard graduate students to work in the Cambridge Public School System. The Division of Engineering and Applied Science (DEAS) has been awarded fellowships from the NSF for Ph.D. students who are committed to working with science teachers and students in the Cambridge Public School System.

The fellows are paired with teachers in Cambridge Rindge and Latin (the local public high school) and with their partner teachers. They work on curricular activity development, laboratory development, and teacher and student support and enrichment. The fellows work with high school students of all ages and grade levels in physics, chemistry, biology, and computer science.

Curriculum Resources[37]

Cambridge Rindge and Latin uses the *Teaching for Understanding* framework for curriculum development. Thus K–12 fellows work to create curriculum modules in their subject areas that work within the framework. 30 lessons were posted for teachers in a variety of subjects.

Massachusetts Institute of Technology–MIT Open Courseware Projects

Highlights for High School Courses[38]

Highlights for High School organizes thousands of MIT introductory course materials into a format that is more accessible for high school students and teachers. The site also features more than 2,600 video and audio clips, animations, lecture notes, and assignments from actual MIT courses and aligns them to the Advanced Placement (AP) topics in biology, calculus, and physics. Courses also include test questions and lecture notes to help students prepare for all assessments.

Teachers Introduction Video[39]

An introductory video that explains how the courses are designed for easy navigation and usage. There is no registration; anyone can use the site including parents that home school.

High School Courses Developed by MIT[40]

The courses listed on the site are from MIT's Educational Studies Program (ESP), an MIT student group offering high school courses for over 50 years and the Chandra Astrophysics Institute.

Other MIT Resources for High School—BLOSSOMS[41]

Blended Learning Open Source Science or Math Studies (BLOSSOMS) is a free repository of video modules for high school math and science classes. The ultimate goal is to gather a large body of content generated by gifted volunteer teachers from around the world.

Resources currently available on the site were provided by MIT faculty and Ph.D. students and include resources in mathematics, physics, and biology. The Web site is growing, and over the coming months there will be BLOSSOMS video modules created by educators in Jordan and Pakistan—all available in English.

MIT Professor Teaches Physics ... His Way[42]

MIT professor and Web star Walter Lewin swings from pendulums and faces down wrecking balls to show students the zany beauty of science.

MIT Video Lecture Series Physics[43]

Professor Walter Lewin demonstrates that the period of a pendulum is independent of the mass hanging from the pendulum. This demonstration can be viewed on the video of Lecture 10. There are 35 video lectures available.

K–12 Outreach Programs[44]

MIT has a wide variety of K–12 outreach programs available to the public. Costs for the programs vary by sponsor. Links and contact information are available at each site.

Cornell University–Nanoscale Science and Technology Facility (CNF)

The CNF Offers Education Opportunities for Middle and High School Students[45]

CNF offers an introduction to nanofabrication and the possibilities of the nano-world, and, depending on the age group—a clean room tour. With enough lead-time and a little assistance from teachers, we can take students (13 and older) into the clean room, show them around, and let them watch as we etch their school mascot or logo into a wafer for them to take home. Because of the nature of our facility and work, we do not offer programs for youths under 9 years old, nor do we take youths under 13 years old into the clean room.

CNF Junior FIRST LEGO® League[46]

The Junior FIRST LEGO® League (Jr. FLL) is geared toward kids ages 6 to 9 and is a spin-off of the FIRST LEGO League (FLL). Each year, the FIRST organization (For Inspiration and Recognition of Science and Technology) releases a science-themed challenge for the teams to work on. In this league for the younger kids, the kids form teams that work on a LEGO building challenge based on this science theme. Teams are formed by teachers, coaches, parents, or other mentors who want to encourage kids in science and technology, with the fun of building with LEGOs. Teams consist of up to six kids, who meet with their mentor to research the topic for that season and build a model based on the specific building challenge given. The kids can be classmates, friends, siblings, neighbors, 4-H clubs, etc. Any group can form a team to participate. The season culminates with the high-energy music-thumping Expo, where the teams gather to present their results to the public and to friendly reviewers, and everyone receives a prize for their work.

CNF FIRST LEGO® League[47]

The FIRST LEGO® League (FLL) is a team-based competition for kids ages 9 to 14 utilizing the LEGO® MINDSTORMS® robotics kits. Each year, the FIRST organization (For Inspiration and Recognition of Science and Technology) releases a science-themed challenge for the teams to work on. The challenges consist of a table-top LEGO® course for the MINDSTORMS® robots to work through, and a research project for the team to investigate, develop ideas, and then present in a creative presentation.

California State Summer School for Mathematics and Science (COSMOS)[48]

Designed specifically for talented and motivated high school students, the California State Summer School for Mathematics and Science (COSMOS) is a four-week summer residential program for high school scholars with demonstrated interest and achievement in math and science.

The program is also open to exceptionally advanced and emotionally mature 8th graders capable of participating in a one-month program away from home. This intensive experience is intended to encourage the brightest and most promising young minds to continue their interest in these fields. Located on four University of California campuses (Davis, Irvine, Santa Cruz, and San Diego), COSMOS provides students with an unparalleled opportunity to work side-by-side with outstanding researchers and university faculty, covering topics that extend beyond the typical high school curriculum.

University of California, Santa Barbara–Materials Research Laboratory

The Education Programs at the Materials Research Laboratory (MRL) provide professional development opportunities for K–12 teachers, research experiences for undergraduates and community college students, and educational experiences for K–12 students.

UCSB ScienceLine[49]

MRL's "Ask-a-Scientist" project enables K–12 students to directly ask expert UCSB researchers their science questions. Topics include astronomy, marine biology, physics, computers, materials science, and earth science, among many others.

MRL Multimedia Highlights[50]

New MRL educational videos explain the excitement and central role of materials science and other science topics. MRL's "Meet a Scientist" program is an extension of ScienceLine, in which elementary school students interview UCSB scientists, and other videos presenting fun topics in materials science.

Sir Harry Kroto Video—Students can watch Nobel Laureate Harry Kroto presenting the structure of the C-60 molecule ("buckyball") to 5th graders.

"Build a Buckyball" Workshop[51]

Students, ranging from elementary to early high school, are given a brief introduction focused on materials science and the scale of the nanometer, different forms of carbon and the relationship between molecular structure and material properties. Afterward, each student builds his own six-inch carbon-60 molecule.

It's a Material World![52]

This program is available for elementary school science nights. Children are naturally curious about science. These hands-on activities inspire inquiry into materials science-related topics and emphasize the fascinating nature of science.

Opportunities for Teachers[53]

Research Experience for Teachers (RET)[54]

Secondary school teachers participate in laboratory research, under the mentorship of graduate, post-doctoral, and faculty researchers. During their second summer in the program, teachers work collaboratively to translate their research experiences into curriculum resources. All of the curriculum projects are archived on the Internet in a database.

Models and Materials[55]

This teacher professional development program brings together teams of art and science teachers from local junior high and high schools to work together on developing curriculum projects that are in line with the state standards for art and science. Models and Materials is a three-year teacher professional development program that brings together teams of local teachers from junior high and high schools. The teachers develop integrated curriculum modules that bring visual art concepts to the science classroom and science concepts to the art classroom. The program introduces materials science, but provides a new way to communicate scientific concepts to students.

Teachers' Projects and Lesson Plans[56]

Modules and lesson plans developed by the teachers are archived on the site.

Annual Curriculum Workshops for Science Teachers[57]

The MRL hosts workshops and symposia on a variety of topics related to science education, outreach, and teacher professional development. The investigative teaching resources developed by teachers who participated in the RET program are presented to secondary science teachers throughout the Santa Barbara County area. All of the curriculum projects are also archived on the Internet in a database.

The 2010 focus is on Enhancing Core Science Curriculum and Presenting Resources for Teaching Climate. The workshop targets upper elementary teachers and secondary science teachers.

In the first part of the workshop, RET teachers present their curriculum models which focus on "Transforming an Earth Science Curriculum" that can be integrated into any middle or high school earth science course; activities focus on the History of Life for middle school biology classes and a unit, "Atoms to Asteroids," that excites students about chemistry and general science.

The second part of the workshop focuses on the important topic of climate change and presents the latest research; it offers information and many hands-on activities and resources for connecting with climate change and addresses ecology impacts of climate change, ocean acidification, and UCSB materials science research on alternative energy solutions.

This center also has technology outreach with industrial partners for workforce training, along with an International Materials Center for global outreach.

Rice University–Houston, TX—K–12 Outreach Programs

The NanoKids™ Educational Outreach Program[58]

Developed by Dr. James M. Tour, Chaos Professor of Chemistry and Director of the Carbon Nanotechnology Laboratory at Rice University. The program is dedicated to increasing public knowledge of the nanoscale world and the emerging molecular research and technology that is rapidly expanding internationally. The animated videos are based on actual anthropomorphic molecules synthesized in the laboratory. Dr. Tour's development of the NanoKids™ visual concept utilizes universally recognized forms exhibiting human characteristics to instruct, motivate, and entertain.

Tested in the Houston Independent School District, they significantly increase students' comprehension of chemistry, physics, biology, and materials science at the molecular level. They provide teachers with conceptual tools to teach nanoscale science and emerging molecular technology. The program demonstrates that art and science can combine to facilitate learning for students with diverse learning styles and interests. In 2006, they added a Hispanic Community Network[59] to the program, expanding the outreach to a larger student population. The network has access to videos, workbooks, sample tests, and teacher resources.

Teacher Resources[60]

Introduction to Program, Card Game of Periodic Table, Answer Key, Links to YouTube Videos

Video Resource[61]

Introduction to NanoKids Series, Structure of Matter, NanoKids Video Music, NanoKids Chapters 1–4.

NanoKids Workbook[62]

Interactive workbook in English and Spanish

Sample Texts[63]

Sample tests for teachers with answer key

Rice University—Center for Biological and Environmental Nanotechnology (CBEN)

Education Outreach Programs[64]

The CBEN established goal of educational outreach and human resource programs is to cultivate a future workforce experienced in using science and engineering at the nanoscale to solve problems in biological and environmental engineering. CBEN's educational outreach activities are coordinated by Dr. John Hutchinson (jshutch@rice.edu), Director for Education, and Dr. Carolyn Nichol (cnichol@rice.edu), Associate Director for Education. CBEN faculty members and students contribute substantially to these programs.

Teacher Professional Development

Nanotechnology for Teachers Chem 570[65]

The Center for Biological and Environmental Nanotechnology (CBEN) at Rice University offers a spring course on Nanotechnology for chemistry, physics, and IPC teachers. The program is designed for high school science teachers, and middle school teachers are welcome. Content in Nanotechnology for Teachers aligns with many of the Texas Essential Knowledge and Skills for Science (TEKS).

Course features:

- **IPC:** Motion, waves, energy transformations, properties of matter, changes in matter
- **Biology:** Structures and functions of cells and viruses; cells, tissues, and organs; nucleic acids and genetics
- **Environmental Systems:** Interrelationships among resources and environmental systems
- **Chemistry:** Characteristics of matter; energy transformations during physical and chemical changes; atomic structure; periodic table of elements
- **Physics**: Conservation of energy and momentum; force; thermodynamics; characteristics and behavior of waves; and quantum physics

Nanotechnology Internship for Teachers[66]

CBEN offers a 4-week internship in research laboratories, which is focused on techniques for translating the center's research methods and results into a classroom setting. Piloted in the summer of 2002, the program reinforces a teacher's sense of science as a process, and deepens the understanding of achievements and potential of nanoscience and the ability to apply lessons learned from the content class to a research setting. The program requires development of inquiry-based lesson plans that apply CBEN research, use of digital media to bring laboratory experience to students, reading technical articles related to research topics, and networking with fellow science teachers.

Resource for Teachers[67]

Webcasts of Chem 570, Nano Lecture materials, videos of lab demos, and lesson plans developed by the teachers for middle and high school.

K–12 Student Enrichment Programs[68]

CBEN has three summer enrichment programs to encourage high school students with varied social and economic backgrounds to study science and move toward science and engineering majors and careers in nanotechnology. The Nanoscience Discovery Academy is open to 11th and 12th grade high school students from any school or school district. The JPMorgan Chase Career and Knowledge Institute and the Schlumberger Nanochemistry Academy are partnerships between CBEN and Project GRAD (Graduation Really Achieves Dreams) and recruit 9th grade students from Yates, Wheatley, Sam Houston, Davis, and Reagan High Schools.

Undergraduate Programs[69]

CBEN's undergraduate educational development programs are designed to engage a diverse population of students in nanotechnology and to effectively prepare them for rewarding careers in the field of nanotechnology. The program offers Nanotechnology Research Experience for Community College Students, sponsors undergraduate outreach opportunities with inner city high school students, and develops new courses in nanotechnology, while augmenting existing courses. Nanotechnology Research Experience for Community College Students conducts research in a Rice University lab under the guidance of leading faculty and graduate students. Undergraduate Curriculum CBEN faculty offer courses for students interested in nanosciences.

Graduate Programs[70]

Graduate Student Interviews—Meet some of Rice's current graduate students and learn about their research in the nanosciences. Student Leadership Council CBEN graduate students have self-assembled into a new Student Leadership Council with committees on research, communications, education, and careers. Graduate Curriculum CBEN faculty offer a variety of courses in the nanosciences.

University of Virginia, Charles L. Brown Department of Electrical and Computer Engineering UVA Virtual Lab K–12[71]

Developed from an NSF grant in 2005, under the direction of John C. Bean, J. M. Money Professor, the "UVA Virtual Lab" features the Nanosurf® EasyScan desktop scanner from NanoScience Instruments. The science education Web site has detailed instructions for teachers and students on the use of microscopy that eliminates the need for separate in-person teacher training. The clarity of the design and tutorials brings microelectronics, nanotechnology, and the underlying science to college, precollege students, and the general public. It replaces math and difficult concepts with intuitive 3D animations. Microelectronics presentations explain how semiconductors and transistors work and how they are fabricated in both university labs and factories. Nanoscience presentations describe alternate forms of nanocarbon, the process of DNA self-assembly, and the inner workings of instruments used to see at the nanoscale (such as SEMs, AFMs, and STMs). These pages link back to basic science presentations on electricity, magnetism, and electrical circuits, including "X-ray vision" simulations of common classroom experiments and apparatus. Overall, the Web site contains over fifty presentations on micro- and nanoscience, each illustrated with dozens of virtual reality animations.

Hands-On Introduction to Nanoscience[72] Class Web Site

Under NSF sponsorship, this class was developed to introduce early undergraduates to nanoscience and nanotechnology. The theme: In nanoscience, Newton's sensible laws are replaced by the weirdness of quantum mechanics. The consequences: First, electrons begin to act like waves—but because all waves are similar—experiments with light and water waves offer insights into electron behavior. Second, at the nanoscale one can no longer use light-image-based microfabrication to make things directly. Instead one has to design the parts so he or she knows how he or she wants them to finally come together (the ultimate example of this self-assembly—DNA synthesis of protein). And finally, to confirm that things worked the way they were planned, the need for new instrumentation to see things at the nanoscale (such as scanning tunneling and atomic force microscopes). The class Web site provides a full set of PowerPoint lecture notes covering these topics (including figures, animations, readings, and lists of demonstration equipment). It also includes full guides to student laboratory use of miniaturized STMs and AFMs. Both lectures and labs make use of 3D animations provided by the sister "UVA Virtual Lab" Web site focusing on microelectronics, nanotechnology, and their underlying science.

Nanoscience Lessons by Teachers for Elementary and Middle Schools

It is fairly easy to repackage the class material for high school students. But can nanoscience also be brought into primary and middle school grades? The developers explored that challenge in compressed (one week long) summer versions of the nanoscience class taught for Virginia K–12 science teachers. During that week the teachers attended all of the normal class lectures, discussions, and labs. But their homework (completed during the fall) was to then come up with a nanoscience lesson plan for their own students.

Colorado State University–NSF Extreme Ultraviolet (EUV) Engineering Research Center (ERC)

K–12 Outreach[73]

The EUV Center has initiated its outreach to K–12 with programs for students and teachers in Colorado and California. The goal was to make a sustained multiyear effort that had a lasting impact on the intended audience. They offered a variety of opportunities for students and teachers, led by faculty and graduate students and outreach coordinators at the education partnership institutions.

Research Experience for Teachers (RET)

A six-week, hands on, paid research experience for middle school teachers at one of the three NSF ERC EUV sites. During this period, teachers performed research, participated in education workshops on optics, developed new curriculum based on their experience, became the audience for lectures given by professors from the sites, and presented their results at a symposium for all three sites.

Light and Optics Workshops for Teachers

These half-day workshops are held at a local site and involve highly interactive experiments, demonstrations, and inquiry-based discussions on the topics of light and optics. There is a version for middle school teachers and another for high school teachers. The demonstration-question-experiment process advocated in the workshops is designed to elicit student discovery by encouraging a constructivist approach to learning.

The intended goal is to provide teachers with a complete sequence of questions and discussion points, which, in combination with experiments, can successfully steer students to discover for themselves the nature of physical phenomena in light/optics. The actual lab materials that accompany such an effort are often minimal and quite simple. Yet, in the context of a well-designed thought process, they are also very powerful. Thus the emphasis is on the architecture of the discussion rather than the equipment. Each teacher is provided with a box of materials at the end of the workshop.

Light and Optics Lab for Students

On an annual basis, the EUV Center hosts a group of 24 high school students from Poudre High School in Fort Collins, CO. Students attend four separate sessions (held over two weeks) in a teaching lab at the Engineering Research Center on the CSU campus. In groups of two or three, students conduct a series of experiments on optical tables using lab equipment. Students work through each lab guided by a lab manual consisting of a series of questions designed to teach students about reflection, refraction, diffraction, lenses, lasers, fiber optics, polarization, etc. These workshops have been very successful and highly popular with the students.

CU Wizards[74]

Dr. Margaret Murnane created a presentation for the CU Wizards outreach program focused on UV light. The presentation involves a PowerPoint slide show and many hands-on activities that allow people of all ages to appreciate the unique characteristics of UV light.

Now in its 33rd year, CU Wizards is happy to share a new season as they explore the exciting worlds of physics, chemistry, biology, and astronomy. FREE to the public, monthly shows entertain and inform children about the wonders of science. Shows are presented on Saturday mornings once a month and provide a perfect start to a fun-filled weekend.

Let's Make Light

Dr. Carmen Menoni developed a workshop for female elementary school students in Fort Collins, CO. The workshop addresses the properties of light and how it is generated in incandescent and florescent bulbs, LEDs, and light sticks. Elementary school girls participate in the workshop by performing several experiments with these different sources of light. The workshop is part of an annual month-long series aimed at motivating women to enter science.

The Nature of Light

The CU Science Discovery Program[75] is currently offering monthly teacher workshops (called Science Explorers) on light that were developed as part of our collaboration with this program. The Education Director, Kristi Dahl, worked with the previous ERC EUV director, Michael Celaya, to develop these hands-on, inquiry-based workshops that have been attended by over 50 teachers and 250 students this year. The CU Science Discovery Program also offers classes and workshops for both teachers and students during the school year and summer on a variety of topics.

University of Colorado at Boulder–Renewable and Sustainable Energy Institute

Science Discovery Program and Science Explorers

In 1988 Science Discovery initiated the Science From CU program of classroom presentations to students statewide. Science Discovery serves over 30,000 students and 1,500 teachers statewide.

Curriculum K–12[76]

Physics for Fun—Grades 4–8[77]

You can provide students with a state-of-the-art physics lab without leaving the classroom. This Emmy Award-winning program is designed for grades four to eight, but can be adapted for older and younger students. Physics for Fun units can be used in science kits, to complement another science curriculum, or as self-contained teaching units.

Little Shop of Physics[78]

The Little Shop of Physics is a group of science educators and science students. They travel the region with a van full of hands-on experiments that have been shared with over 250,000 students. They do not show students science, they help them do science. The goal is to teach people that science is something anyone can do. They also work with teachers through the teacher workshops[79] and the community through the *Everyday Science* TV show.[80]

NASA Nanotechnology Education Outreach[81]

The NASA Nanotechnology Outreach at NASA Ames features an image gallery, archived presentations, reports, a webcast, and a series of movie clips that can be used in the classroom.

NASA Quest Nanotechnology for Kids from NASA Ames[82]

NASA Quest lists all resources by grade level and connects K–12 classrooms with NASA people, research, and science via mission-based challenges and explorations. Join Quest for live interactive webcasts, chats, forums, and online publishing of student work.

The Biologically Inspired Materials Institute (BIMat)

BIMat was established by NASA under the University Research, Engineering and Technology Institute (URETI) program in August 2002. The principal goal for BIMat researchers is to develop bio-nanotechnology materials and structures for aerospace vehicles. The team combines the talents from five of the nation's leading research universities and institutions to advance biomolecular and biomimetic materials design, synthesis, and processing.

NASA's vision: Future-generation space systems will be complex, intelligent, and thinking systems that can adapt, form, evolve, and generally deal with changes and unanticipated problems. High levels of capability will be achieved through highly integrated sensors, articulators, and distributed computer power. Reliability will derive from adaptability and self-repair, just as in biological systems. To be cost-effective, these systems must be designed to minimize weight; weight is often the overriding cost issue in space missions.

Outreach Resources[83]

NASA Educational Television: The NASA SCI Files

The Case of the Great Space Exploration Video

The Tree House Detectives dial up Ms. Dana Novak at the University of California at Santa Barbara to learn how mussels and bloodworms are helping scientists invent new self-healing materials for spacecraft. Seminars, Lectures and Short Courses, Academic Courses, and BiMat Summer Internships are available.

NASA Ames Research Center[84]

Fly by Math and Line Up with Math

Web site invites students in Grades 5–9 to use hands-on math to avoid air traffic conflicts. Interactive student/teacher resources.

NASA Education Resources—Math

NASA JSC–Johnson Space Center—Learning Technologies—Grades 6–12[85]

MathTrax is a graphing tool for middle school and high school students to graph equations, physics simulations, or plot data files. The graphs have descriptions and sound so you can hear and read about the graph. Blind and low-vision users can access visual math data and graph or experiment with equations and datasets.

NASA JSC–Johnson Space Center—Learning Technologies[86]

Where Innovation meets the classroom—programs and resources for grades 6–12.

Calculus Animations, Graphics, and Lecture Notes[87]

PreCalculus through elementary differential equations lessons are provided. The discussion of numerical integration techniques and numerical methods used in elementary differential equations are on the Computer Lab page. All animations and graphics were developed using MathCAD. The lectures were also developed in MathCAD and saved as rich text format files or PDF files. The site contains over 450 downloads, of which over 275 are animations. There are 37 pages covering a range of topics from trigonometry and precalculus through the entire calculus sequence and elementary differential equations.

The Web site was developed by Kelly Laikos, teacher of calculus for 20 years. A small annual donation gives students access for a year for less than the cost of a half-hour session with a tutor.

Interactive Mathematics Links[88]

This site provides links for high school math that are interactive and are usually written in Java. Java is a programming language that allows the user to input data, control diagrams by dragging and rotating, and even play games. Most browsers accept Java, so you do not need to worry. However, some of the sites require Shockwave, which is a plug-in. You will need to follow the directions for installing Shockwave if it is indicated.

University of Illinois–Center for Nanoscale Chemical-Electrical-Mechanical Manufacturing (Nano-CEMMS)

One of the primary missions of the Nano-CEMMS Center is to develop a diverse workforce of educators, scientists, engineers, and practitioners to advance nanomanufacturing technology in the United States and around the world. All graduate students, postdoctoral associates and fellows, researchers, and professors affiliated with the Center assist in this educational mission.

High School Student Outreach[89]

Nano-CEMMS educators provide in-school programs that inform students about a variety of nanotechnology-related topics and future career opportunities in nanotechnology. Programs can be customized for various classroom types, ages of the students, and time frames available.

Nano-CEMMS also conducts NanoChallenge, an after-school program for high school students in the East Central Illinois area. The Center also participates in the Principal's Scholars program by presenting nanotechnology modules within the engineering strand of the program.

Programs for K–12 Teachers[90]

The Center has two primary ways by which K–12 teachers can interact:

First, the Center, in collaboration with Champaign Community Unit School District #4 and Education for Employment System #330, runs the Eastern Central Illinois Nanotechnology Teacher Enhancement Program (ECINTEP). This program offers a two-week summer institute; teachers learn how scientists and engineers work to manipulate matter at the molecular level. Scientists at the University of Illinois and in industry demonstrate current research and the practical applications of research principles in product development. All material developed links to the Illinois Learning Standards, the PSAE Assessment Frameworks, and Advanced Placement courses. You can arrange a single-day, hands-on workshop for up to thirty elementary, middle, and high school teachers. These workshops offer teachers the opportunity to introduce topics related to the Center's research efforts and are usually conducted during in-service days.

Curriculum Modules[91]

A variety of teaching modules are available online. Some of these modules describe labs that can be performed in the classroom by students using the procedure described and accompanied with video files demonstrating each stage of the module. Others come with a kit providing all the necessary tools and materials to complete the module that are shipped to you at no cost. Check out the catalog of modules online.

Stanford University and IBM–Center for Probing the Nanoscale

Stanford University and IBM Corporation, with funding from the National Science Foundation, have founded the Center for Probing the Nanoscale to achieve five principal goals.

1. Develop novel probes that dramatically improve our capability to observe, manipulate, and control nanoscale objects and phenomena.
2. Educate the next generation of scientists and engineers regarding the theory and practice of these probes.
3. Apply these novel probes to answer fundamental questions and to shed light on technologically relevant issues.
4. Disseminate our knowledge and transfer our technology, so that research scientists and engineers can make use of our advances, and corporations can manufacture and market our novel probes.
5. Inspire thousands of middle school students by training their teachers at a Summer Institute.

Education programs span teacher development, workshops, public nanoscience events, academic courses, and graduate education. CPN seeks to embrace the cultural, gender, racial, and ethnic diversity of the United States and is dedicated to increasing the participation of underrepresented groups in science, technology, engineering, and mathematics (STEM).

Outreach Programs

NanoTeachers is a teacher resource for bringing nanoscience and nanotechnology into the classroom.

Even though the subjects of nanoscience and technology are not part of the curriculum, the exciting science and technology can be brought into the classroom through various standards (California Science Standards related to nanoscience and nanotechnology).

Summer Institute for Middle School Teachers[92] was created to foster middle school students' interest in science through teacher training. The one-week Summer Institute for Middle School Teachers (SIMST) is scheduled annually. At the institute teachers learn about the physical concepts and underlying concept of nanotechnology and nanoscience in simple terms. Daily sessions focus on content lectures and inquiry-based modules that explicitly address California's 5th to 8th grade physical science content standards. Teachers also receive a hands-on activity classroom kit with many fun activities that bring nanoscience into the classroom.

Hands-On Nano Activities[93] Grades 2–12

Hands-on activities are listed online and sorted by grade level for teachers. These activities, originally developed for teachers participating in CPN's Summer Institute for Middle School Teachers, are designed to help students learn about nanotechnology while supporting the California Science Content Standards. For teachers in the San Francisco Bay Area, CPN has also partnered with the nonprofit Resource Area for Teachers (RAFT) to make available low-cost materials for some of these activities.

Partnership with the National Hispanic University, to pique students' interest in nano science.

Postdoctoral Mentoring Program, to support CPN postdocs to follow through with their careers.

Probing the Nanoscale, a graduate level course that teaches the underlying concepts of nanoprobes and showcases their applications.

Annual Nanoprobes Workshop brings together industry and academia to foster collaboration as well as inspire and inform participants.

Video Education Resources[94]

To promote nanoscience education, the Center for Probing the Nanoscale offers free videos of many Center classes and programs. In addition, the CPN has partnered with Stanford's Office of Accessible Education on the Proteus Project and hopes to make many course materials available in alternative formats including searchable captioned videos and transcripts of the lectures.

University of California, Berkeley–Center of Integrated Nanomechanical Systems (COINS)

The goal of COINS is to develop and integrate cutting-edge nanotechnologies into a versatile platform with various ultrasensitive, ultraselective, self-powering, mobile, wireless communicating detection applications. The success requires new advances in nano-electro-mechanical devices, from fundamental building blocks to enabling technologies to full device integration. The Center developed four major research programs in the areas of energy, sensing, mobility, and electronics/wireless. Each program research project spans the full spectrum of basic through applied level with a set of criteria to determine the projects to support assuring optimal project alignment.

Nano Camp for High School[95]

COINS has partnered with UC Berkeley's Lawrence Hall of Science to provide a summertime opportunity for high school students for hands-on experience with nano science and technology. Presented for the first time in 2007, Nano Camp offers two one-week, nonresidential sessions at the Lawrence Hall of Science. Scholarships are widely available.

Students work in teams to synthesize a ferrofluid (a magnetic fluid with unusual properties first developed by NASA), etch electrode structures using the photolithographic technique, fabricate a photovoltaic cell using nanoparticles, and more. The program is complemented by tours of labs and interactions with Berkeley NSE scientists.

Nanotechnology Workshop for Teachers[96]

Building on the success of the Nanotechnology Camp for High School Students, the one-day workshop introduces high school and middle school teachers to basic concepts and current research in nanoscience and nanotechnology. Teachers come away prepared to engage their students in three labs: self-assembly demonstration, ferrofluid, and solar cell. They receive a $200 stipend and a kit that includes: binder with activities, articles, and materials resource information and class set of materials and chemicals for activities done in the workshop. Recruitment for participation targets teachers in underresourced schools.

Summer Math and Science Honors (SMASH) Academy[97]

The goal of the Summer Math and Science Honors (SMASH) Academy is to encourage students from underrepresented communities (Hispanic/Latino, African American, and Native Americans) to pursue studies and excel in science, technology, engineering, or math at top colleges and graduate schools.

Beginning in the summer after ninth grade, SMASH participants spend five weeks on the UC Berkeley campus attending science, math, and English classes, taught by experienced high-school teachers.

Proven successful, the data is impressive. In 2008, 96 scholars from over 55 Bay Area high schools participated in SMASH, with approximately 90% from underresourced schools. In 2007, 100% (18) of the first graduating class are attending college (of which 2 are at UC Berkeley, 2 at community colleges, and the rest at other CSUs and UCs).

Berkeley Nanotechnology Club (BNC)[98]

UC Berkeley is positioned as a leader in nanoscale science and engineering. The BNC was created as the go-to information source for Berkeley students and alumni to connect with nanotechnology resources at Berkeley and the Bay Area. The goals are to solidify Berkeley's position as a nanotechnology leader and foster an interdisciplinary community that promotes knowledge sharing and nano-entrepreneurship.

Membership is free and ensures that you are kept up-to-date with everything nano at Berkeley.

Northeastern University (NEU)–Center for High-Rate Nanomanufacturing (CHN)

The Center for High-Rate Nanomanufacturing is focused on developing tools and processes that will enable high-rate/high-volume bottom-up, precise, parallel assembly of nano-elements (such as carbon nanotubes, nanoparticles, etc.) and polymer nanostructures. CHN projects fall into three categories: large scale directed assembly and transfer research, environmental health and safety research, and regulatory and ethical research, in addition to education and outreach programming.

Middle School and High School Outreach[99]

CHN researchers are happy to travel to schools and present their work to students of all ages and host groups that wish to tour the facilities at NEU, University of Massachusetts–Lowell (UML), and University of New Hampshire (UNH). The laboratory facilities in NEU's Kostas Nanomanufacturing Center and UML's Nanomanufacturing Facilities and Plastics Processing Laboratories receive a constant stream of visitors. K–12 students touring the UML's facilities usually check out the electromicroscopes and injection mold parts, produce blow-molded bottles, electrospin polymer solutions, and examine current research projects. Students touring the Kostas Nanomanufacturing Center cannot usually enter the clean rooms, but undergraduate students prepare posters explaining the processes that the students view through the windows of the clean rooms. In addition, longer programs allow high school students to fabricate parts.

Research Experiences for Teachers[100]

Teachers are recruited for six-week research experiences for a teacher (RET) program that is overseen by Claire Duggan at NEU's Center for Enhancement of Science and Mathematics Education (CESAME). Participants are placed in research assignments in pairs at NEU and UML and work a minimum of 35 hours per week at their sites. Most of their time is spent in the research laboratory working on some aspect of CHN research. As part of the research experience, all participants are required to keep a laboratory research notebook, a reflective journal, and develop a research presentation to be shared with colleagues and affiliates at the end of the summer.

University of Nebraska-Lincoln–Materials Research Science and Engineering Center (MRSEC)

K–12 Outreach[101]

The goal of MRSEC outreach efforts is to inspire kindergarten to grade 12 (K–12) students by the excitement of materials research to enlarge the pool of students who become future scientists and engineers. Outreach efforts take various forms, such as classroom visits and bringing K–12 students to the University of Nebraska. Visits to schools by MRSEC researchers provide hands-on science activities and demonstrate to students the relevance of materials and nanoscience research. Bringing groups of K–12 students to the University of Nebraska for tours and demonstrations of the MRSEC laboratories informs them about scientific concepts and new technologies. This provides the opportunity for students to learn about materials and nanoscience research career paths. A few examples of K–12 outreach activities are as follows.

How Strong Is It? First Grade

University of Nebraska MRSEC faculty worked with first-grade students at Morley Elementary School for seven science lessons to study many properties of magnetism and magnetic materials. These enthusiastic students used hands-on activities to learn how to tell whether a material is magnetic, how to make magnets in a variety of ways, how the earth behaves as a giant magnet, and how like poles repel and unlike poles attract. They were especially intrigued by the question of how "strong" a magnet could be and delighted in testing to find out (January 2009).

Fourth Graders Study Optical Properties of Solids

MRSEC researchers visited Clinton Elementary School to help students investigate the optical and electrical properties of solids. Twenty-three fourth-grade students did several different hands-on experiments involving electrical measurements, optical reflection and transmission, and observations of the refractive properties of transparent material. The students made and tested electrical circuits to study the conduction of various materials, including aluminum and glass. They used light from low-power lasers and other sources to study reflective and refractive properties of materials and how materials are shaped so that images can be formed. The students also investigated the colors of clothes observed through different optical filters. The students learned that electrical conductors usually reflect light better than nonconductors and that some materials transmit light quite well. The session ended with a very active and wide-ranging question and answer period where such matters as the properties of light from the sun and how different materials are made were discussed.

Academy Day at North Star High School

What could be more entertaining than fire and good music? How about fire that will dance to that music? The Rubens' tube, also known as the standing wave flame tube, is a physics experiment demonstrating a standing wave. It shows the relationship between sound waves and air pressure and also allows you to see different wavelengths of audio frequency.

This and other experiments were demonstrated by MRSEC faculty member Christian Binek during two fifty-minute presentations to high school students during Academy Day at North Star High School. These presentations reached approximately 500 high school students who are considering various careers in science or technology (April 2009).

MRSEC Collaboration with Raymond Central High School

MRSEC researcher Jeff Shield visited three ninth-grade science classes and Pam Rasmussen's senior physics class at Raymond Central High School. The ninth-grade students were introduced to the magnetic properties of materials and their applications in magnetic information storage on hard disk drives. These demonstrations were directly related to Mrs. Rasmussen's summer research with MRSEC. The senior physics students worked in small groups to measure the electrical properties of light emitting diodes. They also used diffraction gratings to study the properties of the emitted light.

Seventh Graders Make Nanowires

Nebraska MRSEC researchers Bernard Doudin and Carolina Ilie visited Scott Middle School in Lincoln to discuss advances in nanotechnology. The students made their own batches of silver nanowires with diameters ranging from 30 to 80 nm by depositing Ag in polycarbonate membranes. They learned how to estimate the wire diameter from the color of the membrane. They also showed that a single membrane of 1 cm diameter contained more wires than the number of people in the United States.

Materials Science for Elementary/Secondary Students

University of Nebraska MRSEC faculty made presentations on materials science topics to students in grades 5 to 12 classrooms and in larger venues such as "Future Problem Solvers" Workshops and "Saturday Science." Presentations include such topics as magnetic materials, shape memory alloys, optics and light, and nanotechnology. The discussion often includes answers to the questions: "What does a scientist do? How do I become a scientist?"

RET Participant Uses MRSEC to Increase Student Interest in Science

An important outcome of Research Experiences for Teachers (RET) programs is helping teachers communicate the excitement and relevance of materials research to their students. Steve Wignall, a teacher from Seward High School who worked with MRSEC Professor Diandra Leslie-Pelecky on biomedical applications of magnetic nanoparticles, brought fifteen students to visit the University of Nebraska. In addition to visiting labs, students learned about materials research career paths. Many were surprised to learn that materials research involves a range of scientific and mathematical backgrounds and has applications from information storage to medicine. Wignall reports that students returned with a new appreciation for materials research and increased excitement about math and science classes. Plans are to expand the program in the future by involving students from other RET-associated high schools and offering additional information on math and science careers.

The University of Alabama–Center for Materials and Information Technology

High School Internship Program[102]

The Center for Materials for Information Technology at the University of Alabama invites high school sophomores and juniors to apply for a summer research internship in nanoscience and engineering. The students will join the nanoscience and engineering research team and do basic research on either new materials for high-density information storage, new catalysts for energy storage and conversion, or nanomedicine. Faculty from chemistry, physics, mathematics, chemical engineering, electrical engineering, and metallurgical and materials engineering form multidisciplinary research teams. They solve fundamental materials science problems that allow the information storage industry to continue to make data storage devices (magnetic hard disk drives, magnetic tape, MRAM) with higher data storage capacity at lower cost. The students find rewarding careers as scientists and engineers in the high-tech information storage industry. High school sophomores and juniors who are highly motivated toward a career in research in science and engineering are invited to apply each year.

University of Maryland (UMD)–Materials Research Science and Engineering Center (MRSEC)

Project Lead the Way[103] (PLTW)

The UMD-MRSEC has formed a six-year partnership with Charles Herbert Flowers High School Project Lead the Way (PLTW) to implement its pre-engineering courses. PLTW is a pre-engineering program that introduces students to the scope, rigor, and discipline of engineering and engineering technology prior to entering college. Major goals of the program are to increase the number of young people who pursue engineering and engineering technology, provide equitable and inclusive opportunities for all academically qualified students without regard to gender or ethnic origin, and contribute to the continuance of America's national prosperity.

MRSEC provides students with classroom inquiry-based activities, hands-on university-based instruction days, and summer programs. The MRSEC program includes:

Instructional topics: Materials science and engineering, engineering design and production, electronic circuits, and nanotechnology.

Laboratory visits: Modern engineering materials instructional lab, nanofabrication lab, nanoscale imaging, spectroscopy, properties lab, and Keck Laboratory for Combinatorial Nanosynthesis and Multiscale Characterization.

Chemistry Program[104]

MRSEC established a partnership with Oxon Hill High School to provide students with a chemistry program designed by teachers and MRSEC members. The goal was to increase students' general and AP chemistry knowledge and experimental skills. The program included classroom visits and field trips to the University of Maryland. During classroom visits, MRSEC researchers present exciting demonstrations to motivate students, followed by laboratory experiments. The field trip is an extensive program, which provides students the opportunity to use university laboratory facilities and advanced laboratory equipment. The program also includes career talks by researchers and graduate students, department of chemistry and biochemistry admissions staff, and a discussion with students about undergraduate experiences.

MRSEC-AIP Middle School Student Science Conference[105]

The Annual Middle School Student Science Conference is cohosted by the Materials Research Science and Engineering Center (MRSEC) and the American Institute of Physics (AIP) Education Division. The conference has gained wide recognition and interest from area schools and the community. This program is designed for middle school and draws students from diverse backgrounds. The conference program is guided by three objectives: (1) to improve students' abilities in and perceptions of science, (2) to improve students' research presentation skills, and (3) to involve families and community in science education. To accomplish these objectives, three components have been developed: (1) mentors work with students to carry out the scientific method by completing a science fair project; (2) mentors work with students to help transform their science fair projects into scientific presentations in one-on-one mentoring sessions; and (3) students present their projects at the conference day event.

Presentations are judged based on content, style, and how well students address questions from the judges and audience. The audience consists of parents, teachers, mentors, and distinguished community members.

Girls Excelling in Math and Science (GEMS)[106]

The UMD-MRSEC, in collaboration with Girls Excelling in Math and Science (GEMS), implements math, science, and engineering activities to middle school students. GEMS is an after-school program focused on female students from two Prince George's County Middle Schools. The MRSEC-GEMS program is designed to enhance these students' interests in science and engineering by providing exciting curriculum, effective studying tools, science demonstrations, and access to cutting-edge laboratory equipment. MRSEC researchers interact with students as mentors and role models while providing students with the opportunity to learn and discuss current research and technology.

Summer Camps[107]

The UMD-MRSEC summer camps were established in 2001. The camps are designed with teachers to meet curriculum needs and standards and to build appreciation for science and engineering. Students from partner schools and home school programs attend. High school camps include specific curriculum goals that are incorporated into students' fall semester learning and final semester evaluation. The camps also strive to build confidence in students' abilities in and perception of science and engineering. The camps provide a venue to share cutting-edge science and technology and to generate interest among budding young scientists.

Science Road Shows[108]

The UMD-MRSEC takes science road shows to partner schools in order to stimulate and improve science learning by emphasizing fundamental science concepts and building science literacy using more than 50 different demonstrations. Working with teachers, UMD-MRSEC tailor-design shows to accompany conceptual units, laboratories, and science fair development, and introduce new concepts, motivate learning, and engage critical thinking. UMD-MRSEC also presents demonstrations in public venues to inspire interest in science and technology among the broader community. Topics include physics, materials science, and nanoscience. Venues include Six Flags, partner schools, summer camps, professional workshops, conferences, and the International Spy Museum.

Florida State University–Molecular Expressions: Exploring the World of Microscopy[109]

The Molecular Expressions Web site features the photo galleries that explore the fascinating world of optical microscopy. It includes the largest collections of color photographs taken through an optical microscope (commonly referred to as "photo-micro-graphs"). Students and teachers can visit the Photo Gallery for an introductory selection of images covering just about everything from beer and ice cream to integrated circuits and ceramic superconductors. These photographs are available for licensing to commercial, private, and nonprofit institutions.

Secret Worlds: The Universe Within[110]

Soar through space starting at 10 million light years away from the Milky Way down to a single proton in Florida in decreasing orders of magnitude (powers of ten). This tutorial explores the use of exponential notation to understand and compare the size of things in our world and the universe, and provides a glimpse of the duality between the macroworld around us and the hidden microworld within. As you move through space toward the Earth in successive orders of magnitude, you reach a tall oak tree just outside the buildings of the National High Magnetic Field Laboratory in Tallahassee, Florida. After that, you begin to move from the actual size of a leaf into a microscopic world that reveals (100 nanometers) leaf cell walls, the cell nucleus, chromatin, DNA, and finally, into the subatomic universe of electrons and protons.

Online Activity Guidebook for Teachers[111]

Activities written by educators with input from scientists, researchers, students, and teachers are provided here to help you and your students investigate light, optics, and color. Each activity has a teacher version followed by a separate student page. However, student pages can be easily adapted for use by children working with their parents, instead of a teacher. In fact, students, teachers, and parents are encouraged to work together to cooperatively explore new ideas, concepts, and ways of thinking. Then, after students achieve a better understanding of the basic principles and theories behind optics and related topics, they can explore them to advanced levels through the various features of the Molecular Expressions Web site.

Activities for Students[112]

This section has 16 activities and 10 inquiries for students. The photo galleries are also an area of interest to students of any age.

Scanning Electron Microscopy[113]

Award-winning electron microscopist Dennis Kunkel produced a virtual Scanning Electron Microscope (vSEM). Visitors can adjust the focus, contrast, and magnification of microscopic creatures and samples of dust and pollen viewed at thousands of times their actual size.

The NIEHS Kids' Pages[114]

"We Are the Environment." Charles Panati

These pages are produced by the National Institute of Environmental Health Sciences, of the National Institutes of Health, Department of Health and Human Services.

Topics on site: NIEHS ... What's that? (Learn about NIEHS), Hot Topics and Environmental Health Science Education, Brainteasers and Riddles, Books and Stories (educational too!), Environmental Art and Poetry Gallery, Games and Surprises, Jokes and Humor (Laughing is Good for You!), Sing-Along with NIEHS.

Science Education Resources for Kids and Teachers[115]

The Environmental Health Sciences Education Web site provides educators, students and scientists with easy access to reliable tools, resources and classroom materials. It seeks to invest in the future of environmental health science by increasing awareness of the link between the environment and human health.

References*

1. http://www.nano.gov/html/edu/home_edu.html

* All links active as of August 2010.

2. http://www.nano.gov/html/centers/nnicenters.html
3. http://ice.chem.wisc.edu/index.html
4. http://www.mrsec.wisc.edu/edetc/cineplex/index.html
5. http://mrsec.wisc.edu/Edetc/index2.html
6. http://mrsec.wisc.edu/Edetc/supplies/nanoventure/index.html
7. http://www.materialsworldmodules.org/
8. http://www.materialsworldmodules.org/mwmnet/mwmnet.shtml.
9. http://www.nclt.us/
10. http://www.nanoed.org/nlr/Introduction_of_Emerging_Science_into_the_Classroom/
11. http://www.nanoed.org/courses/courses_by_level/100001.
12. http://www.nanoed.org/concepts_apps/nanocos/
13. http://nsf.gov/funding/pgm_summ.jsp?pims_id=5465&org= NSF&from=fund
14. http://cltnet.org/cltnet/index.jsp
15. http://www.nsec.ohio-state.edu/outreach_Ia.html
16. http://cnse.albany.edu/
17. http://cnse.albany.edu/research/profiles.html
18. http://cnse.albany.edu/Nano_for_Kids.html
19. http://cnse.albany.edu/Nano_for_Kids/nano_games.html
20. http://www.cise.columbia.edu/mrsec/
21. http://www.lrsm.upenn.edu/outreach/ec4hsst09.html
22. http://www.lehigh.edu/~inimagin/
23. http://invsee.asu.edu
24. http://invsee.asu.edu/invsee/invsee.htm
25. http://gk12.asu.edu/
26. http://gk12.asu.edu/physical+sciences
27. http://gk12.asu.edu/node/7
28. http://gk12.asu.edu/lessons
29. http://www.mirc.gatech.edu/?q=content/instructional_units
30. http://www.mirc.gatech.edu/?q=content/nanocamp
31. http://www.mirc.gatech.edu/?q=content/nanooze-exhibit
32. http://www.mirc.gatech.edu/?q=content/ret_program
33. http://www.mirc.gatech.edu/?q=content/reu_program
34. http://www.mirc.gatech.edu/?q=content/nanofans-forum
35. http://www.mirc.gatech.edu/?q=content/nanotech
36. http://nanohub.org/tags/k12/resources
37. http://gk12.harvard.edu/
38. http://ocw.mit.edu/high-school/introductory-mit-courses/
39. http://ocw.mit.edu/high-school/for-teachers/
40. http://ocw.mit.edu/high-school/courses/
41. http://ocw.mit.edu/high-school/other-mit-resources-for-high-school/
42. http://thoughtware.tv/videos/show/1618
43. http://ocw.mit.edu/OcwWeb/Physics/8-01Physics-IFall1999/VideoLectures/index.htm
44. http://web.mit.edu/K–12 edu/list.html.
45. http://www.cnf.cornell.edu/cnf5_eduoutreach.html
46. http://www.cnf.cornell.edu/cnf_jrfll.html
47. http://www.cnf.cornell.edu/cnf_fll.html
48. http://epc.ucsc.edu/cosmos/index.shtml

49. http://www.scienceline.ucsb.edu/
50. http://www.mrl.ucsb.edu/mrl/outreach/educational/multimed3.html
51. http://www.mrl.ucsb.edu/mrl/outreach/educational/buckyball.html
52. http://www.mrl.ucsb.edu/mrl/outreach/educational/materialworld.html
53. http://www.mrl.ucsb.edu/mrl/outreach/educational/opportunities.html
54. http://www.mrl.ucsb.edu/mrl/outreach/educational/RET/index.html
55. http://www.mrl.ucsb.edu/mrl/outreach/educational/models/index.html
56. http://www.mrl.ucsb.edu/mrl/outreach/educational/models/curriculum/pages.html
57. http://www.mrl.ucsb.edu/mrl/outreach/educational/workshops/workshops.html
58. http://nanokids.rice.edu/
59. http://cohesion.rice.edu/naturalsciences/nanokids/mission.cfm?doc_id=8402
60. http://nanokids.rice.edu/resources.cfm
61. http://nanokids.rice.edu/videos.cfm
62. http://nanokids.rice.edu/workbook/
63. http://nanokids.rice.edu/sampletests.cfm
64. http://cben.rice.edu/education.aspx
65. http://cben.rice.edu/CHEM570.aspx
66. http://cben.rice.edu/teacherinternship.aspx
67. http://cben.rice.edu/education/resources.aspx
68. http://cben.rice.edu/K–12 programs.aspx
69. http://cben.rice.edu/undergradprograms.aspx
70. http://cben.rice.edu/graduateprograms.aspx
71. http://www.virlab.virginia.edu
72. http://www.virlab.virginia.edu/Nanoscience_class/Nanoscience_class.htm
73. http://euverc.colostate.edu/education/k12.shtml
74. http://www.colorado.edu/physics/Web/wizards/cuwizards.html
75. http://www.colorado.edu/ScienceDiscovery/
76. http://www.colorado.edu/sciencediscovery/teachers/curriculum.html
77. http://www.colorado.edu/sciencediscovery/teachers/physics4fun.html
78. http://littleshop.physics.colostate.edu/
79. http://littleshop.physics.colostate.edu/workshops.html
80. http://littleshop.physics.colostate.edu/tvshow.html
81. http://ipt.arc.nasa.gov/gallery.html
82. http://quest.nasa.gov/projects/nanotechnology/resources.html
83. http://bimat.princeton.edu/html/outreach.html
84. http://smartskies.nasa.gov/
85. http://prime.jsc.nasa.gov/mathtrax
86. http://prime.jsc.nasa.gov/
87. http://calculus7.com/
88. http://www.fcps.edu/DIS/OHSICS/math/socha/index.html
89. http://www.nano-cemms.uiuc.edu/programs/high_school
90. http://www.nano-cemms.uiuc.edu/programs/k_12_teachers
91. http://www.nano-cemms.uiuc.edu/materials
92. http://www.stanford.edu/group/cpn/education/simst.html
93. http://www.stanford.edu/group/cpn/teachers/activities/
94. http://www.stanford.edu/group/cpn/education/video.html

95. http://mint.physics.berkeley.edu/coins/index.php?option=com_content&task=view&id=101&Itemid=53
96. http://mint.physics.berkeley.edu/coins/index.php?option=com_content&task=view&id=102&Itemid=53
97. http://mint.physics.berkeley.edu/coins/index.php?option=com_content&task=view&id=103&Itemid=53
98. http://www.ocf.berkeley.edu/~nano/
99. http://www.northeastern.edu/chn/education_outreach/middle_school/
100. http://www.northeastern.edu/chn/education_outreach/research_experiences/
101. http://www.mrsec.unl.edu/education/K–12 .shtml
102. http://www.mint.ua.edu/education/highschool.asp
103. http://mrsec.umd.edu/Education/PreCollege/PLTW.html
104. http://mrsec.umd.edu/Education/PreCollege/Chemistry.html
105. http://mrsec.umd.edu/Education/PreCollege/SSC.html
106. http://mrsec.umd.edu/Education/PreCollege/GEMS.html
107. http://mrsec.umd.edu/Education/PreCollege/Summer/
108. http://mrsec.umd.edu/Education/PreCollege/Road.html
109. http://micro.magnet.fsu.edu/index.html
110. http://micro.magnet.fsu.edu/primer/java/scienceopticsu/powersof10/index.html
111. http://micro.magnet.fsu.edu/optics/activities/teachers/index.html
112. http://micro.magnet.fsu.edu/optics/activities/students/index.html
113. http://micro.magnet.fsu.edu/primer/java/electronmicroscopy/magnify1/index.html
114. http://www.niehs.nih.gov/kids/home.htm
115. http://www.niehs.nih.gov/health/scied/

10

Overviews of Nanotechnology Workforce Programs

> There are no such things as applied sciences, only applications of science.
>
> **Louis Pasteur (1822–1895)**

Most nanotechnology workforce training programs are being developed as two-year community college or technical college programs, or a combination of both. Very few states have addressed workforce education of technicians that addresses the needs of companies moving past start-up phase into full manufacturing. This involves stage one, composites and materials for enhancement of products in the marketplace, to stage two, devices and fabrication of products for manufacturing.

Primary and secondary science educators have not introduced nanoscale science in public schools, nor trained the teachers to implement an expanded curriculum base of physics, chemistry, and biology in high school. However, current advanced placement (AP) science courses that high school students complete in these subjects could qualify them for nanotechnology workforce training programs. The new nanotechnology businesses are the result of multidisciplinary fields of discovery in universities and NSF-funded nano centers. Scientists working in physics, chemistry, biology, engineering, information technology, metrology, and other fields are contributing to today's research breakthroughs that require new technicians as commercialization expands globally.

The worldwide workforce necessary to support the field of nanotechnology commercialization is estimated at 2 million by 2015. How does the U.S. educational system train these workers, and how do students choose the appropriate educational path for their interests? As in other fields, a passion for science is developed while students are young, and an introduction to the many facets of nanotechnology will provide the basis for future educational opportunities.

Industry Needs for Nanotechnology Education

Mr. Russell Maguire, Boeing, and Dr. Celia Merzbacher, Semiconductor Research Center (SRC) presented a second area of workforce education at the *"Partnership for Nanotechnology Education Workshop,"* held in 2009. Boeing discovered they needed to retrain their engineers as polymeric composites recently replaced metal in the aerospace sector.

Example of an Industry Approach to Reeducating Its Workforce in Nanotechnology[1]

> When the Boeing 787 all-composite aircraft was launched, it spurred many engineers with decades of metallurgy experience to seek training in polymeric composites technology. Boeing, in conjunction with the University of Washington, developed a multi-course, multi-layer, university-level curriculum to address this enormous demand. The result was a composite certificate program for professional engineers that won national awards and still is in great demand. In the face of an almost equal hunger among these same engineers for skills and knowledge in nanotechnology Boeing is now planning a similarly styled cross-cutting curriculum, to be kicked off in Fall 2010 and involving several leading universities.

The workshop report of the presentation contributed by Dr. James Murday defined a few more issues that need to be addressed:

> Despite a lack of quantitative data, industry workforce needs are varied, growing, and expected to change through time. At the outset, an emerging technology such as nanotechnology likely will impact industry at the research level and companies will seek scientists and engineers with advanced degrees to introduce and incorporate new concepts into products and services. Very quickly, as technologies are being developed for a particular application, companies of all sizes need technician-level workers with appropriate skills. Even among advanced degree holders, the needs of industry are diverse. Small companies in particular may seek employees with a breadth of education, obtained through multidisciplinary degree programs or an emphasis or major in more than one field. Such workers are better able to shift as needed. Large companies with thousands of technology workers may seek employees with specific skills that depend on the sector. Small and large companies may refresh knowledge and skills of current employees through short courses, seminars by outside experts, workshops, etc. Some states offer support for such activities.
>
> Early applications of nanotechnology are almost all based on novel materials and their properties. As a result, industries need workers with knowledge and skills in materials science. At the upperclass and advanced degree level, universities could help by incorporating coursework in nanomaterials science into almost all physical science and engineering programs. At the technician level, there is a need for 1 and 2 year programs that teach students specific skills for synthesis and/or characterization of nanomaterials. And industrial hygienists need specialized (refresher) training to ensure workplace safety.

> The needs of industry for high quality, well-educated scientists and engineers depend heavily on the larger education ecosystem. To ensure a robust supply of workers, the entire ecosystem must be healthy, beginning with K–12 math and science education and continuing through various forms of postsecondary education. Rather than a one-way path from K–12 to undergraduate to graduate school, society can be better served by interactions and partnerships among industry, academia, government, and the broader formal and informal education system.

Meet a Pioneer in Developing a NanoEducation Statewide Program

While I was reviewing the outreach programs across the United States, it was apparent that the work of Professor Steven J. Fonash at Pennsylvania State University (Penn State) exemplified the importance of early planning and statewide inclusion of all stakeholders in education, government, and industry. As an early adopter of these concepts, Dr. Fonash realized in 1992 that the Nanofabrication Facility could anchor a statewide program. He worked tirelessly to bring these stakeholders together to prepare students in Pennsylvania for a workforce trained as technicians, working with engineers and scientists in nanotechnology commercialization. The resulting Capstone Semester of hands-on experience at the Nanofabrication Facility, started in 1998, has proven to nanotechnology businesses across Pennsylvania that technicians trained in the program are excellent employees. Proven to be reliable self-starters, they are flexible and jump right into any position required on the job. Since the program is so successful, it has been placed in this chapter as a model for other states to follow as they struggle to develop expanded K–12, workforce training and expansion of 2-year and 4-year degree programs.

Following the Pennsylvania State programs are overviews of programs developed at Texas State Technical College, Dakota County Community College, The College of Nanoscale Science and Engineering (CNSE), State University at Albany, and Foothill College in California. The section closes with a program developed by Walter Trybula, Texas State University and Dominick Fazarro, University of Texas–Tyler.

Pennsylvania State University–Center for Nanotechnology Education and Utilization, and Nanofabrication Facility

The first statewide nanotechnology educational initiative that included: K–12, Community Colleges, Vocational and Tech Schools, Workforce Training, and a university-wide nanotechnology minor degree was initiated in 1998.

Dr. Stephen J. Fonash had the foresight to develop a state-wide Nanotechnology Education Partnership with a university-wide nanotechnology minor at Penn State University. This early program, initiated in 1998, enables two-year degree and four-year degree students attending various institutions across Pennsylvania to obtain an education in nanotechnology that prepares them for the workforce. The partnership includes Penn State's University Park Campus, 26 other institutions, Pennsylvania industry, and the State of Pennsylvania, offering a total of 54 two-year and four-year nanotechnology degrees each year. The common element in all the two-year and four-year degree programs: students must spend one semester at Penn State attending the hands-on nanotechnology fabrication, synthesis, and characterization immersion. The six-course hands-on experience titled the *Capstone Semester* exposes students to state-of-the art equipment and cleanroom facilities provided at the Nanofabrication Facility, located at Penn State's University Park Campus.

The 18 credits of coursework can be used toward an associate or baccalaureate degree, to earn an NMT Certificate, or both, depending on the specific program of the student's "home institution." Refinement of the Capstone Semester is carried out in close consultation with industry members of the NMT Program's Advisory Board. Two-year degrees include nanobiotechnology, nanomanufacturing, and chemical technology. Four-year degrees include physics, chemistry, biology, and engineering, all with a concentrated focus in nanotechnology.

The Capstone Semester Video[2]

To understand the importance of the Capstone Semester and the career paths for the future, this short video is exemplary.

Careers in Nanotechnology Information Video[2]

This introduction video is extremely informative for anyone interested in developing programs or looking at career options for the future.

K–12 Services and Resources[3]

The Outreach for K–12 Office at Pennsylvania College of Technology provides services and resources to teachers, counselors, administrators, and students in K–12 schools. Annual activities of the Outreach for K–12 Office: Career Day (Fall and Spring), Educator and Counselor Mini Conference, Gold Medal Initiative, Real Game Training, Cake and Pastry Display Show, Food Show.

Outreach for K–12 activities fully support the Pennsylvania State Academic Standards.

Nanotech Summer Camps[4]

In 1999, the NMT Partnership began offering three-day summer "Nanotech Camps" for high school students from across Pennsylvania. These nanotech camps provide secondary school students with an orientation to basic nanofabrication processes and applications and the opportunity to observe these same nanofabrication processes in the Penn State Nanofabrication Facility.

Nanotech One-Day Camps

One-day Nanotech Camps are offered year round by appointment. These camps include an overview of nanotechnology and a tour of the Penn State Nanofabrication Facility.

Degrees That Work Video[5]

This 30-minute presentation explores nanotechnology and the related career opportunities with experts in the field. It follows a former Penn College student, Mark A. Atwater, as he completes his education at one of the nation's top nanotechnology education facilities, the Penn State University's Center for Nanotechnology Education and Utilization.

Nanofabrication Workshops for Educators[6]

These workshops explore the world of nanotechnology and are offered several times per year. Participants learn about the growing applications of nanotechnology in many industries including the biotechnology, MEMS, optoelectronics, chemical, and nano-electronics industries. The fundamentals of nanofabrication processes and tools are covered and practiced through processing labs in the "class 10 cleanrooms" of the Penn State Nanofabrication Facility. Each workshop runs three days with daily morning lectures by Penn State researchers and engineering staff, and afternoon lab sessions in nanofabrication. The intended audiences are high school and college educators, administrators, guidance counselors, and career counselors. A previous knowledge of micro- or nanotechnology is not required. Pennsylvania school teachers attending these workshops may receive Act 48 credits, available upon request.

Education Tools—Activities[7]

Amazing Creatures with Nanoscale Features—Part 1 (K–12)

This animation is an introduction to microscopy, scale, and applications of nanoscale properties. It introduces some of the tools that are used by scientists to visualize samples that are smaller than what we can see with our eyes. This includes the optical microscope, scanning electron microscope, and the atomic force microscope. In this animation, you will take a closer look at a butterfly wing at different magnifications and see features at the nanoscale that give the butterfly unique properties. Then, you learn how scientists and engineers mimic these structures through engineering techniques.

Education Tools—Nanotechnology Video Modules[8] (High School and Workforce)

The video modules are provided as part of the introduction to nanotechnology courses at partner institutions. Modules are educational standpoint and industrial standpoint. Each contains a set of three videos for streaming on the Internet. Students and teachers considering career paths in this new workforce gain an understanding of the benefits that nanotechnologies bring to society through a multitude of product areas. Dr. Fonash and Amy Brunner expertly answer questions for the students who are considering the Capstone Semester program at the Nanofabrication Facility; the Pennsylvania nanotechnology companies that hire graduate technicians also answer student's questions about skill levels necessary to work in the various fields.

Nano4Me[9]

Nano4Me is a community of individuals serving the national nanotechnology education and workforce development initiatives. Nano4Me is maintained by the Penn State College of Engineering with contributions from educators and industry personnel from across the country. This community includes parents, students, program graduates, educators, industry personnel, government officials, and others.

Pennsylvania Partnership

The Penn State College of Engineering has been a national leader in nanotechnology education and workforce development since 1998. The Pennsylvania NMT Partnership was created with support for the Pennsylvania Department of Community and Economic Development, to help meet the growing needs of Pennsylvania industry for skilled nanofabrication workers.

National Connection

From 2001 to 2008 the College was home to the National Science Foundation (NSF) Regional Center for Nanofabrication Manufacturing Education, sponsored by the NSF Advanced Technological Education (ATE) program. In 2008 the National Center for Nanotechnology Applications and Career Knowledge (NACK Center) was established through the NSF ATE program as a successor to the Regional Center.

Partners

Penn State College of Engineering, PA NMT Partnership/CNEU, National Center for Nanotechnology Applications and Career Knowledge (NACK Center), DCED, National Science Foundation (NSF).

The Nanotechnology Workforce Development Initiative (NWDI) Texas

NWDI is a key to a statewide effort to support advanced nano manufacturing in Texas; to stimulate entrepreneurial efforts and attract new business to the state. The Texas Workforce Commission supported the project to supply the workforce for the new class of services and products emerging from the nanotechnology industry.

The participants (Texas State Technical College–Waco, Baylor University, Del Mar College of Corpus Christi, Zyvex Corporation, Sematech, and Richland College) joined forces in the nanotechnology workforce initiative to lay the foundation for an innovative and effective partnership to produce a qualified workforce based on the growing demands of nanotechnology.

The project drew interest from partners in 2004, due to the potential economic impact of the commercialization of research at Zyvex Corporation in Richardson, TX. The company was funded by a five-year $25 million cost share award from the National Institute of Standards and Technology–Advanced Technology Program (NIST-ATP), that supported the commercialization of manufacturing technology at Zyvex, enabling new companies to drive the nanotech applications used in advanced manufacturing.

The Nanotechnology Workforce Development Initiative was funded with a Wagner-Peyser Act grant, administered by the Texas Workforce Commission (TWC). The new program is built around the successful training paradigm already established in TSTC Waco's partnership with Baylor University's Center for Astrophysics, Space Physics and Engineering Research (CASPER). The center's experimental lab is housed on the TSTC Waco campus and allows students from TSTC Waco's laser electro-optics and semiconductor manufacturing programs to serve as interns. The CASPER internship and Sematech site visits and interactions laid the groundwork for more advanced nanotechnology field experiences for students at Zyvex.

Several other key colleges, such as Richland Community College and TSTC Harlingen, along with leading nanotechnology companies such as Intelliepi, Nanotechnologies Inc., and Nanospectra Biosciences, Inc., have joined the partnership. It is anticipated that, as this program evolves, more nanotechnology companies will look to the community colleges to recruit highly trained and talented nanotechnicians.

The long-term goal is to engage the state's 51 community colleges, establishing partnerships and career tracks for their students and graduates and opportunities to include nanotechnology courses in their curricula.

The partnership expects this program to serve as a statewide model for community colleges to adopt. As the commercialization cluster of nanotechnology companies continue to grow, more partnerships are likely to develop to train workforce technicians statewide.

The Texas Industry Cluster Initiative Reports and Assessments[10]

The Texas Industry Cluster Initiative is leading the state to build the future economy by focusing on strengthening competitive advantage through six target industry clusters.

> This cluster initiative is important because for the first time in the history of this state, we will have a coordinated, market-driven economic development strategy that focuses on areas where we have the greatest growth potential and focuses on fostering that potential.
>
> **Governor Rick Perry**

The assessment phase of the initiative engaged a cross-section of over 700 stakeholders throughout the state. Collectively, the findings and recommendations call for strategies to develop a skilled workforce, a competitive education system, and an effective commercialization process, supported by a highly efficient supply chain.

Reports are available on the following topics: Texas Industry Cluster Initiative Background, Acknowledgements, Advanced Technologies and Manufacturing Cluster Assessment, Aerospace and Defense Cluster Assessment, Biotechnology and Life Sciences Cluster Assessment, Information and Computer Technology Cluster Assessment, Petroleum Refining and Chemical Products Cluster Assessment, Energy Cluster Assessment.

Curriculum Syllabus Download[11]

Overview of nanotechnology
Nano measurements
Nano characterization
Nano technology systems

Outreach Program Modules[12]

Twelve modules related to major areas of nanotechnology research and application were developed by very well-respected individuals in their fields. The modules were created in PowerPoint format with accompanying notes provided below each slide to understand the material. Review the *"Introduction to Nanotechnology"* module first if you are a newcomer to the subject. The modules are a public outreach part of the partnership for businesses, high school, middle school, or elementary science teachers, college or university professors, or people who are merely interested in this emerging field.

Dakota County Technical College (DCTC); Rosemount, Minnesota

Deb Newberry, MSc., Contributor

In 2002, two innovative leaders (Vice President Karen Halvorson and Dean Mike Opp) at Dakota County Technical College in Rosemount, Minnesota, started thinking about the possibility of creating a nanoscience technologist program at DCTC. Based on a strong history of customized training courses directed at meeting the needs of area industry and a foundational network of company relationships, DCTC, with approval of President Dr. Rom Thomas, decided to host a Nanotechnology Symposium in February of 2003.

The focus of the symposium was to discuss what was happening in the region with regard to nanotechnology research and applications and to determine at a fundamental level if there was a need for a nano-savvy workforce. Over 50 companies were represented, and the program consisted of area educational and technical leaders—as well as venture capitalists, authors, and industry representatives—involved in an open discussion about the "state of nanotechnology" within Minnesota. At that symposium it was determined that there was an interest and a need for a trained nanoscience workforce. Dr. Halvorson asked for volunteers to work on an advisory committee with DCTC in a detailed need assessment and the creation of a nanoscience program at Dakota County Technical College.

Hence, a 2-year NanoScience Technologist Program which has become a model for other 2-year programs around the country was born. The above review of the history of the creation of the program was discussed in order to emphasize the involvement, dependence, and correlation with industry and workforce needs that has been the cornerstone of the DCTC program.

Shortly after the February 2003 meeting, industry focus groups were held. These focus groups addressed two questions: (1) Is there a true need for a nano-savvy workforce within the Minneapolis/St. Paul region? And (2) if so, what are the knowledge, skill, and ability requirements for these workers? There was a representation from a large variety of companies—which drove the curriculum content—and included 3M, Honeywell, Medtronic, General Mills, Entegris, Hysitron, Smart Skin Technologies, Lloyds Foods, HB Fuller, Valspar Paints, and Surmodics.

The answer to the first question was a resounding "yes"—there was definitely a need for technician-grade employees that understood nanotechnology. In many cases companies were using masters and Ph.D.-educated employees to perform technician-grade jobs mainly because they were the only employees who knew how to do nano-related tasks.

The answer from industry to the second question defined the details of the four-semester program that exists at DCTC today. Figure 10.1 represents a summary of "what was heard" and some of the challenges to be faced. An overall challenge for the creation of the DCTC NanoScience Technologist program was the four-segmented aspect of what industry needed. In addition to needing employees that knew how to operate,

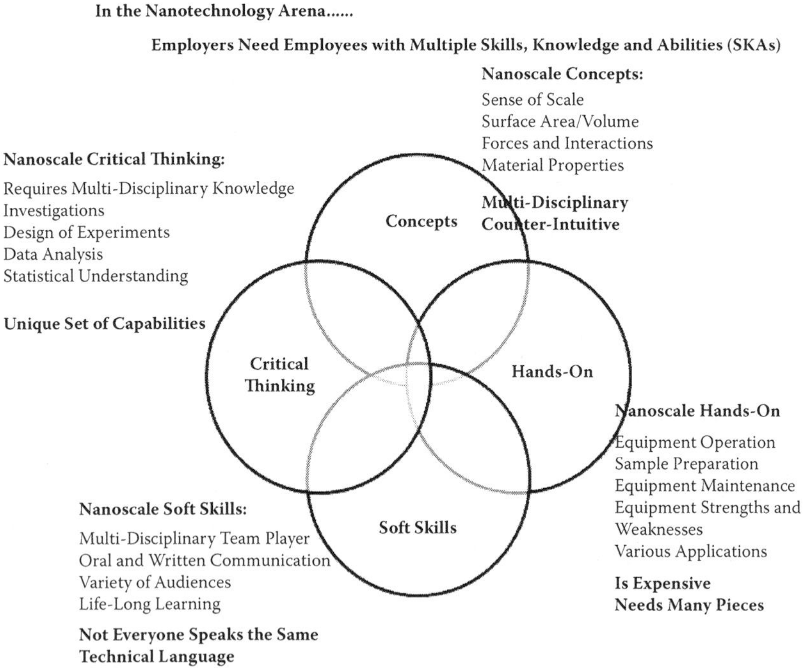

In the Nanotechnology Arena......

Employers Need Employees with Multiple Skills, Knowledge and Abilities (SKAs)

Nanoscale Concepts:
Sense of Scale
Surface Area/Volume
Forces and Interactions
Material Properties

Multi-Disciplinary
Counter-Intuitive

Nanoscale Critical Thinking:
Requires Multi-Disciplinary Knowledge
Investigations
Design of Experiments
Data Analysis
Statistical Understanding

Unique Set of Capabilities

Concepts
Critical Thinking
Hands-On
Soft Skills

Nanoscale Hands-On
Equipment Operation
Sample Preparation
Equipment Maintenance
Equipment Strengths and Weaknesses
Various Applications

Is Expensive
Needs Many Pieces

Nanoscale Soft Skills:
Multi-Disciplinary Team Player
Oral and Written Communication
Variety of Audiences
Life-Long Learning

Not Everyone Speaks the Same Technical Language

FIGURE 10.1
SKA.

maintain, and prepare samples for the various tools of nanoscience (atomic force microscopes, scanning electron microscopes, etc.), the companies wanted employees who understood the concepts of the traditional sciences at the nanoscale. They also wanted employees who could think on their feet, think "outside of the box," and so on. Finally, these future employees needed to be able to communicate, work well in teams, and analyze and present data.

Employers wanted these technicians not to be "same process every day" technicians, but they wanted technicians who could actively participate in product research and development teams and understand (to a certain degree) the science and technical aspects behind what they were doing.

The creation of the DCTC program also faced one more challenge. That challenge was the fact that, within the Twin Cities region, the need for nanoscience technologists was not driven by one particular industry segment. Companies within the twin cities that had a strong interest in nanotechnology included medical research and device companies (biotechnology), electronics companies (traditional and research level devices), and material science (coatings, adhesives, polymer science) as shown by the focus group participants. This meant that the DCTC program needed to be multidisciplinary in nature.

Figure 10.2 is a representation of the integration and overlapping that was required during the curriculum content creation phase. Concepts from traditional sciences that specifically applied to the general concepts of nanotechnology such as surface area to volume ratio and exponents (math), atomic structure and bonding (chemistry), forces and interactions (physics), and protein structure (biology) were addressed with the nanoscale in mind—pulling from the traditional classes but adding in nanoscale relationships.

It should be noted that students from multiple programs at DCTC take many of the general education classes such as physics, chemistry, etc. So the intent was not to modify or perturbate the traditional general education classes, but to work in lock step with those syllabi within the context of the nano-specific classes.

During the initial organization of contextual topics and courses, industry researchers were continuously involved. The researchers help assess the content, approve the arrangement of the topics, and also influence the level of detail as well as drive the specific skills, knowledge, and abilities that would comprise the outcome of the courses and the curriculum. Industry representatives were also involved in the creation of a comprehensive set of competencies for the entire NanoScience Technologist Program.

Finally, Dakota County Technical College did not have any of the nano-specific lab equipment required for the students to achieve the level of hands-on expertise that was required by our industry customers.

For this purpose DCTC and the University of Minnesota created a partnership by which the students would attend DCTC for the first three semesters of the program and then attend the University of Minnesota

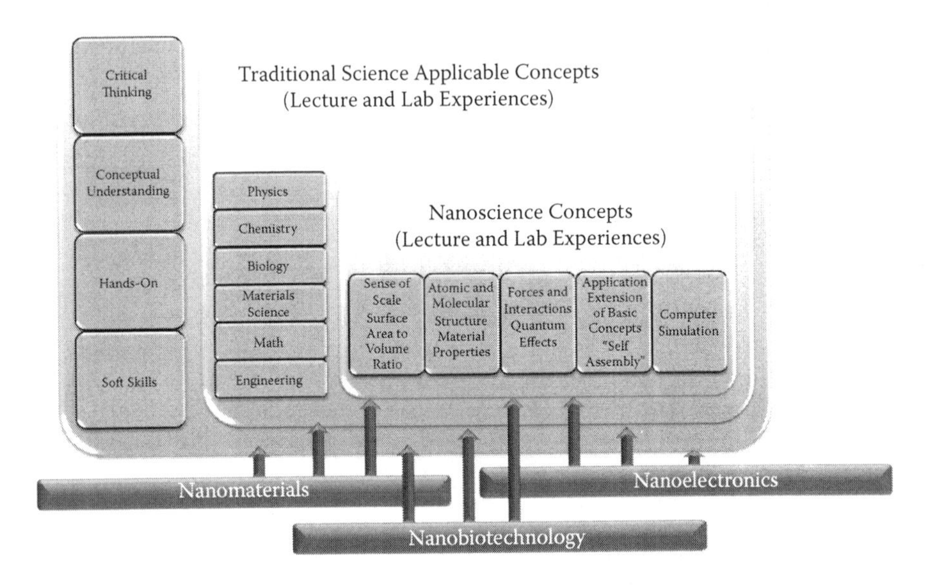

FIGURE 10.2
DCTC nanoscience program.

for a fourth semester lecture and lab experience. Dr. Steve Campbell was a significant influence in this partnership and was (and continues to be) instrumental in coordination of content and activities between the two schools.

DCTC applied for and received a National Science Foundation Grant in 2004. This grant provided for the finalization of the curriculum content and initial program offering. The resulting 72 credit AAS Degree program is shown in Figure 10.3. Note that nanotechnology content is offered in each semester, allowing students to gradually increase their depth of understanding and application of the concepts of nanoscience. Also note that half of the program credits are for nano-specific content. Finally, students are required to do a 320-hour minimum participatory internship. For most students the internship is much longer, and they are often hired by the company for which they interned.

In 2006, DCTC graduated their first class of 13 from the AAS Degree Nanoscience Program. Since then approximately 10 students have graduated each year.

The program continues to enjoy strong industry support through internships and jobs for program graduates. The curriculum content is continuously reviewed and enhanced to address the evolving needs of industries.

Dakota County Technical College
Nanoscience Technology Program Course Outline and Credit Allocation
rev. 2009

Semester 1 at DCTC			Semester 2 at DCTC			Semester 3 at DCTC			Semester 4 at Univ. of MN		
Course	Name	Credits	Course	Name	Credits	Course	Name	Credits	Course	Name	Credits
BIOL 1500	General Biology	4	CHEM 1500	Introduction to Chemistry	4	NANO 2101	Nano Electronics	3	MT 3111	Elem. of Micro Manufacturing	3
PHYS 1100	College Physics I	4	PHYS 1200	College Physics II	4	NANO 2111	Nanobiotechnology/ Agriculture	3	MT 3112	Elem. of Micro Mfg Lab	1
COML 1400	Intro to Computers	3				NANO 2121	Nanomaterials	3	MT 3121	Thin Films Deposition	3
ENGL 1100	Writing & Research Skills	3	SPEE 1020	Interpersonal Communication	3	NANO 2131	Manufacturing, Quality Assurance	2	MT 3131	Intro to Materials Characterization	3
MATS 1300	College Algebra	4	MATS 1250	Principles of Statistical Analysis	4	NANO 2140	Interdisciplinary Lab	3	MT 3132	Materials Characterization Lab	1
NANO 1100	Fund. of Nano I	3	NANO 1200	Fund of Nano II	3	NANO 2151	Career Planning and Industry Tours	1	MT 3141	Principles and Applications of Bionanotechnology	3
			NANO 1210	Computer Simulation	1				MT 3142	Nanoparticles & Biotechnology Lab	1
									NANO 2970	Internship	2
Credits		13 to 21	Credits		19	Credits		15	Credits		17

DAKOTA COUNTY
TECHNICAL COLLEGE

FIGURE 10.3
DCTC course outline.

Companies provide tours, specific experiments, and research projects for students and often appear as guest lecturers emphasizing the application of the knowledge in addition to the research aspect.

In 2008, DCTC was awarded a National Science Foundation grant for Nano-Link. Nano-Link is a Regional Center for Nanotechnology Education. Partner institutions reside within North Dakota, Minnesota, Wisconsin, Illinois, and Michigan. We continue to emphasize nanotechnology education with a direct correlation to industry needs. Nano-Link will continue to expand nanotechnology training for two-year institutions as well as high school educator and student outreach.

An Introduction Video[13] of the program and overview are available on the Internet.

A Myriad of Opportunities

There are a myriad of opportunities available for those who teach and study nanotechnology—a variety of skills levels, a diverse range of industry options and career directions, as well as a chance to combine many aspects of science. These opportunities also represent challenges especially for educators and workforce training development.

These challenges include (1) providing hands-on experience early in the training or program to improve student retention, (2) having access to the expensive nanotechnology tools, (3) being able to explain intangible concepts with concrete examples at the nanoscale, and (4) providing a platform for investigative and critical thinking experiments or activities.

In 2008 the leadership at NanoInk, a nanotechnology company in Skokie, Illinois, decided to create a nanotechnology education system based on their commercial equipment. NanoInk designs and sells dip-pen lithography equipment. Based on a nanoscale approach similar to dipping a quill pen in an inkwell and writing with it, NanoInk has created tools to create patterns at the nanoscale level. These patterns can have a variety of applications from microfluidic devices, DNA, toxin and protein sorting and detection, to electrical circuits and so on—it is a multi-application piece of equipment.

Using the dip pen lithography equipment as the focal point, NanoInk added an atomic force microscope and a fluorescent microscope to complete the suite of equipment. They then added an extensive text curriculum and eight laboratory experiments to create the NanoProfessor[14] Nano 101 Course. The initial user group for this equipment is 2-year colleges, with extensions to high schools also planned. In December of 2009, DCTC became the first college in the nation to have the suite of equipment and curriculum and is serving as the "pilot" program for the NanoProfessor, and the experiments and curriculum content will be integrated into the existing DCTC program.

The equipment is versatile and easy to use and addresses all of the challenges that nanotechnology educators face. This complete package of equipment, hands-on lab experiments, and textbook will serve traditional college educators, along with workforce development training needs.

The strength of Dakota County Technical College's NanoScience program is based on a combination of many factors. Partnerships and relationships with regional industry have been a critical aspect of the program since its inception. Continuing to meet the needs of industry will continue to be the top priority for the DCTC program.

The College of Nanoscale Science and Engineering (CNSE), State University at Albany[15]

This is the world's first college dedicated to research, development, education, and deployment in the emerging disciplines of nanoscience, nanoengineering, nanobioscience, and nanoeconomics. With more than $5.5 billion in public and private investments, CNSE's Albany NanoTech Complex has attracted over 250 global corporate partners.

Students, faculty, and corporate partners use leading-edge tools at CNSE's Albany NanoTech Complex, which houses research and development, and prototype manufacturing infrastructure, Class 1–capable clean rooms, and the most advanced 200mm/300mm wafer facilities in the academic world.

CNSE students take part in a groundbreaking course of study. The NanoCollege's unique, cross-disciplinary curriculum, offering B.S., M.S., M.S.-M.B.A., and Ph.D. degree programs, allows students to delve deeply into a wide variety of advanced topics needed to support cutting-edge research.

CNSE's Albany NanoTech Complex serves as an acceleration facility for more than 250 global corporate partners, helping them overcome technical, market, and business development barriers to successfully deploy nanotechnology-based products.

High school seniors[16] conducting research at CNSE receive recognition in prestigious Intel Science Talent Search. From clean energy initiatives to revolutionary health-related research, three high school students took their science projects to the next level under the guidance of CNSE faculty as part of internships at CNSE's world-class Albany NanoTech Complex.

New Bachelor's Degree Program

The bachelor's degree in nanoscale engineering at CNSE complements the undergraduate degree in nanoscale science—launched last June as the world's first comprehensive baccalaureate degree program in nanoscale science—as well as CNSE's graduate-level curriculum, which was also a global first when it commenced in September 2004.

Students completing the CNSE bachelor's degree program in nanoscale engineering will be well-equipped to solve challenging problems in nanoscale engineering and applied sciences, as well as other physical and biological engineering sciences, and be uniquely educated to pursue opportunities in emerging high-technology industries, including nanoelectronics, nanomedicine, health sciences, and sustainable energy, as well as competitive graduate degrees in emerging engineering fields.

The nanoscale engineering curriculum comprises a cutting-edge, inherently interdisciplinary, academic program centered on scholarly excellence, educational quality, and technical and pedagogical innovation. The program will give students the abilities to creatively solve nanoscale engineering problems through the use of rigorous analytical, computational, modeling, and experimental tools based on foundational physical or biological sciences and mathematics, nanoscale materials, and applied engineering concepts.

The program will also enable students to propose, formulate, and execute original, cutting-edge nanoscale engineering research at CNSE's Albany NanoTech Complex. UAlbany NanoCollege offers students a world-class experience working with, and learning from, the top innovative minds in the industrial and academic worlds.

The CNSE houses the only fully integrated, 300mm wafer, computer chip pilot prototyping and demonstration line within 80,000 square feet of Class 1–capable clean rooms. More than 2,500 scientists, researchers, engineers, students, and faculty work on site at CNSE's Albany NanoTech, from companies including IBM, AMD, GlobalFoundries, SEMATECH, Toshiba, Applied Materials, Tokyo Electron, ASML, Novellus Systems, Vistec Lithography, and Atotech.

Resources for Students[17]

NanoCareer Day

NanoCareer Day introduces middle- and high-school students to the exciting world of nanotechnology through tours, presentations, and hands-on activities.

NanoHigh

NanoHigh, developed jointly by the City School District of Albany (CSDA) and CNSE, offers students an unprecedented opportunity to study the emerging field of nanotechnology through two courses that combine classroom learning at Albany High School with monthly on-site visits to CNSE for in-depth lab activities.

NanoEducation Summit

CNSE recently held the region's first-ever NanoEducation Summit for school superintendents, administrators, and teachers, with discussion focusing on ways to implement nanotechnology awareness and learning at the middle and high school level.

Regional Collaboration

CNSE partners with a variety of regional organizations to promote nanotechnology awareness among teachers, students, and parents. Some recent partnerships include:

- Capital District Future City Competition
- Tech Valley Summer Camp
- Girl/Boy Scouts
- Alliance of Technology and Women Great Minds Program

School Presentations and Tours

CNSE hosts a variety of school groups throughout the year—from administrators and faculty to students and their parents. The faculty, staff, and graduate students at CNSE take great pride in providing access to state-of-the-art facilities and a world-class team in an effort to educate the youngest members of our community. CNSE's educational outreach programming inspires students to pursue careers in science and technology, creating the innovative workforce that is critical to our future on a regional, statewide, and national scale.

Foothill College, Los Altos Hills, CA—NSF-ATE Award 0903316 Foothill College Nanotechnician Program[18]

Associate Professor Robert D. Cormia, Contributor

Silicon Valley is home to some of the most innovative research, development, and manufacturing in the world. As the pace of this technology increases, so does the need for skilled technicians who can work productively in R&D, advanced manufacturing, and process characterization. The pressure to emerge from R&D into manufacturing requires skilled technicians who can use sophisticated analytical tools to characterize nanostructures and both develop and optimize new and innovative process engineering. On a daily basis, technicians apply multidisciplinary knowledge of chemistry, physics, engineering, and advanced materials science, to understand and solve critical engineering and manufacturing problems.

Foothill College has developed a specialized program to train technicians with strong material characterization skills using a variety of instruments, and applying scenario-based training to these new and emerging industries. This program provides the knowledge, skills, and competencies needed for the evolving, converging, and emerging technical workplace, using integrated curriculum focused on the application of fundamental nanoscience principles to the design and fabrication of new materials, and materials characterization to support process development. Technicians apply problem solving and QA/QC skills, including image analysis and microscopy, spectroscopy, chromatography, and engineering process measurements, to support high-tech manufacturing. This program addresses a core problem identified in a study by Boeing and SRI (2007)[19] that few college graduates, and even fewer technicians, have the experimental practice and experience to support scientists and engineers in emerging multidisciplinary fields, such as bioengineering and clean energy, and advanced materials engineering and manufacturing. A scenario-based laboratory curriculum ensures that technicians will be productive and effective within demanding environments found in start-up engineering and manufacturing firms and additionally provides a strong multidisciplinary foundation to extend the tenure of their careers.

Our three-year program includes preparatory math and English to address the achievement gap prevalent in California and provide graduates with a much sought after Associate of Science (A.S.) degree in nanoscience technology.

What Questions Are We Addressing?

Future trends and roles of technicians require technician education to stay abreast of rapid advances in the field of materials science and engineering. Technicians today need to work across cross-functional workgroups from R&D to prototyping and early manufacturing. They conduct more experiments, analyze more data, and use more complex data management tools. Technicians in new and emerging multidisciplinary fields, including biomedical materials, clean technology, and advanced (high performance) materials development need a working understanding of chemistry, physics, engineering, and often biology. They work globally, in extended workgroups, use collaborative engineering software, and must perform well in multicultural/multilingual environments. These trends are a result of both globalization and innovation centers which form in multicultural environments.

A nanotechnology industry survey performed by Dakota County Technical College (early 2001) found that technicians need to have an understanding of the nanoscale (nanoscale structure and properties), be able to run instruments and perform simple experiments, and work well in groups. Surveys performed by Foothill College (Nanotechnology 2004 and Clean technology 2008) suggest experience with materials processing and characterization, the ability to support scientists and engineers conducting experiments, and both relevant and specific industry experience were either lacking or sorely needed (Boeing/SRI). This work (Boeing/SRI) found that a majority of engineering graduates lacked relevant exposure to materials development and the nuances of problem solving, and the primary cause of this gap was a dearth of industry (scenario) driven curriculum.

Which Components of Technician Education Programs Work (or Don't Work), with Whom, Why, and under What Circumstances?

Nanotechnology and nanomaterials education has "traditionally" been applied to dedicated silicon technology (wafer and MEMS), and older thin film deposition, which are the workhorses of the electronics industry. More recently nanotechnology has permeated almost all materials-driven technology fields, and especially in the development of high-performance composite materials for transportation, aircraft, and ultralight passenger vehicles. Other notable applications range from thin-film solar photovoltaics (silicon and CIGS), light high-performance batteries (lithium iron phosphate) and new ultra capacitors, to engineered coatings for implantable biomedical devices.

The turnkey training that worked for semiconductors and thin film is not applicable to nanomaterials engineering, which requires application requirements analysis, modeling, material selection and processing, and extensive characterization to develop structure property and process correlations. Additionally, technicians are involved in more problem-solving in rapid prototyping and early manufacturing, and there is no reference manual for the types of problems observed in start-ups.

Which Educational Strategies Have Proven Most Effective in Improving Student Learning in These Specific High-Technology Fields?

Scenario-based learning in conjunction with integrated STEM have proven effective in providing a strong foundation in chemistry, physics, engineering, and biology, with application-centric curriculum providing both relevant context as well as exciting learning opportunities, as shown in Table 10.1.

Can These Strategies Be Translated to Other Fields of Technology (in Addition to Nanotechnology)?

Yes. Foothill College has developed scenario-based learning in robotics and engineering, bioinformatics, computer science, and network security. The PNPA rubric (Figure 10.4) applied to materials science (modeling and design, processing, and characterization) is an effective "systems teaching" tool for integrating cross-functional industry units (R&D, engineering, and manufacturing). Boeing has identified "systems integration" as an industry goal and a likely transition for most major high technology firms employing complex technology, engineering, manufacturing, and support. The newly emerging electric vehicle industry (Silicon Valley and California) and China's channel sourcing and partner engineering are prime examples of this trend. The PNPA rubric can also be decomposed into problem posing, data gathering/experimentation, analysis, and reflection, which are all integral to scenario training.

Nanotechnology programs typically comprise five or more semesters, which include science and engineering courses, and two or more years of mathematics. Within community colleges, these courses are often taught in conjunction with a partnering four-year institution, where expensive instrumentation is housed.

The four-course program described here focuses on a nanomaterials engineering emphasis, starting with an overview of nanotechnology (NANO51) followed by nanomaterials and structures (NANO52), nanomaterials characterization (NANO53), and techniques of nanofabrication (NANO54). These four courses represent the four components of the PNPA rubric, and the materials engineering dogma: applications drive properties (modeling) election, which drives processing. From the resulting structure, properties are characterized, and structure–property relationships are used to add insights into the modeling and fabrication process, enhancing knowledge of novel nanostructures and process optimization.

Additionally, the four-course nanoprogram can also be overlaid on PNPA, as shown in Figure 10.4. Characterization is central to the PNPA rubric and can be a major theme in a program, reinforcing all aspects of

TABLE 10.1

SLO Matrix Following the PNPA Rubric Showing Reinforcing/Integration of Programmatic SLOs

Introduction to Nanotechnology (Applications)	*Materials Characterization (Properties)*
• What is nanotechnology? – Nanoscience – Nanoengineering • How is nanotechnology used? – Advanced materials • What are key application areas? – Energy – Water – Food – Medicine – Materials • What are key nanostructures? – Carbon nanotubes – Dendrimers – Quantum dots • What are key nanodevices? – Biomedical stents – Solar panels – Fuel cells • What are key nanomaterials? – Nanofibers – Thin films – Powders	• What properties are desired? – Strength (Instron) – Stiffness (Instron) – Thermal conduction (DMA) – Electromagnetic (spectrometer) • What structures are needed? – XRD – SEM – Raman • What composition is needed? – XRF/WDX – XPS • What chemistry is desired? – FTIR – XPS • What is the layer/interface? – AES/XPS – XRF – FIB/SEM-EDX • Surface finish? – AFM – SEM – Optical microscopy
Nanomaterials and Nanostructures (Structures)	*Nanofabrication (Processing)*
• What structures are needed? – Fibers – Thin films – Powders • What processing is used? – Thermal – Annealing – Vapor deposition – Spin coat • How to characterize? – Image ▪ Optical, AFM, SEM – Surface ▪ AES, XPS, FTIR – Chemistry ▪ FTIR, XPS – Structure ▪ XRD, metallography	• How to make metals alloys – Powder metallurgy – Sintering • How to make thin films – Vacuum deposition ▪ Evaporative ▪ Sputtering ▪ CVD/Plasma – Spin coat • How to make fibers – Quartz furnace ▪ Methane • How to make powders – Powder metallurgy – Sintering • Novel structures – Self assembly – DNA scaffolding

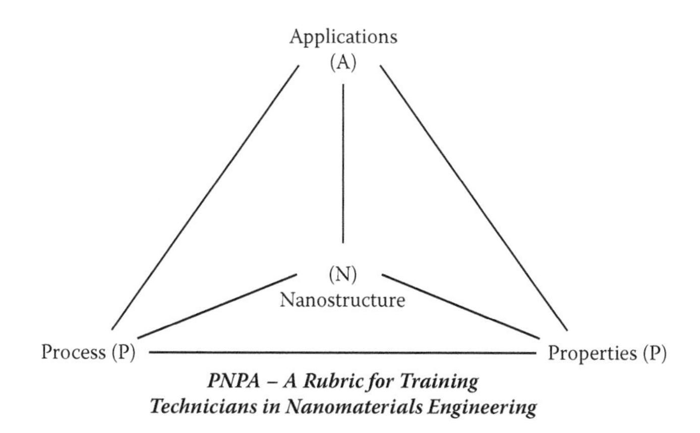

FIGURE 10.4
PNPA.

learning. Additionally, the processing rubric (PNPA) is extensible to any area of high-performance materials and advanced materials engineering and development.

Scenario-based curriculum based on regional industry provides relevant context for both learning and applying engineering skills, and especially for gaining an understanding of industry-specific challenges: PV panels, lightweight batteries, biomedical devices. Emerging technology in clean technology, biomedical devices, and advanced materials each have a focal point. Each has particular devices and materials.

Ten key nanostructures (defined in Table 10.2) represent a fairly complete introduction to the world of nanostructured materials and devices in key emerging technologies. The following list shows ten key nanostructures which are used, as the nexus of the PNPA rubric, in the nanomaterials curriculum:

- Thin-film and amorphous silicon (PV)—solar energy
- Carbon nanotubes (CNT)/composite carbon materials
- Surface coatings and SAMs (self-assembled monolayers)
- Nitinol (biomedical stents)/electropolished alloys—biomedical
- Thin-film and plasma coatings (polyester film)—high-performance glazing
- Particles (coated particles)—biomedicine/powder metallurgy (lithium batteries)
- Dendrimers (nanochemistry)—biomedical drug delivery, synthetic biology
- Polymers and composites/nanoparticle filler—lightweight automotive and aircraft

TABLE 10.2

Nanomaterials, Nanoprocessing, and Nanocharacterization

Nanomaterial	Nanoprocessing/Fabrication and/or Sample Preparation	Nanocharacterization and Properties Measurement
Aerogels*	FIB cross-sectioning of a commercial aerogel for characterization by SEM, TEM	Density, micro-porosity, surface area, and strength to weight ratio
Carbon nanotubes	Fabrication of carbon nanotubes in a quartz furnace using methane	Optical characterization of fibers, XPS characterization of carbon valence band
Catalyst*	Grinding and micro milling of commercial catalysts used in industrial applications	Surface chemistry using XPS, FTIR, and Raman spectroscopy
Composite materials	Cross-sectioning of carbon fiber epoxy materials, stress strain measurements of composite filled polymers	Optical and SEM characterization of cross-section, XPS of surface chemistry/adhesion failure
Dendrimers*	Prepare functionalized dendrimers with reactive end group/derivatives	Surface chemistry and functionalization using FTIR demonstration of C13 NMR
Electropolished coatings	Examine electropolished metals and alloys using various electrochemical and mechanical polishing techniques	Characterize surface chemistry and oxide thickness using AES, XPS, AFM, and SEM techniques
Metals and alloys	Use of heat treatments to control grain size and chemistry, and both micro and nanostructural domains	Characterize grain size using optical and scanning electron microscopy, and AES and XPS
Nanoparticles*	Functionalized nanoparticles with various atomic/molecular coatings (commercial sample)	Determine particle size distribution using optical and SEM techniques
Nitinol shape memory alloy (substrates or frames for biomedical stents)	Electropolishing and mechanical polishing of base metal; Mechanical fatigue cycling and stress–strain	Characterize surface finish with AFM and SEM, and surface chemistry using AES and XPS
Polymeric materials	Addition of composite and or particles to blended polymer, measure strength to failure	Use AFM to characterize surface, XPS/FTIR for surface chemistry
Self-assembled monolayers	Prepare SAMs coating on Au coated glass (commercial kit), and or prepare a multilayer functionalized coating†	Determine coating uniformity, and verify surface functionalization using XPS and FTIR.
Thin films	Prepare various thin film coatings on glass, silicon, and polymer substrates; Prepare samples for characterization using cross sectioning and / or FIB tools (demonstration)	Characterize thin film coating structure, later thickness, and interface chemistry using AES, SEM, XPS, techniques; Measure electromagnetic properties

*optional topic † optional lab

- Silicon materials and Micro Electro Mechanical Systems (MEMS)—LOC, microarrays
- Ceramics, cermat, and electro ceramics/fuel cells—stationary/mobile power, energy

Each structure introduces students to a key industry space, such as energy, and an application and device, such as thin-film solar panels, with an unmet engineering requirement, such as high efficiency, low cost, and durability. For PV energy, thin-film solar utilizes CIGS (Copper Indium Gallium Selenide), amorphous (or monocrystalline) silicon, and a host of new thin-film materials being developed.

Students apply the "PNPA compass" as they navigate the content of the four courses, focusing on Applications in NANO51, Structures in NANO52, Characterization and Properties in NANO53, and Processing and Fabrication in NANO54. A programmatic learning outcome, and differentiator, of the Foothill nanomaterials program is an emphasis on nanostructures as a core competency. Specifically, students who complete these four courses will have a competency in modeling and simulation of properties for existing and future nanostructures,[20] a working knowledge of how to fabricate and characterize them, and a framework for using PNPA in the practice of the engineering method—from requirements gathering through design, fabrication, and developing structure–property and process–structure relationships.

PNPA *is* the engineering process, and especially for developing novel structures to meet the needs of new applications (lightweight composites, high-efficiency thin-film PV, and lightweight high-capacity batteries). In materials engineering programs and course curriculum, PNPA creates a framework from the beginning of the program, which serves as a compass to help students, technicians, and future engineers to understand and apply a methodological approach to intelligent nanomaterials design and selection. Every industry application is dissected into application requirements, desired properties (a requirements document) from which a structure–properties database can be searched and navigated. Existing and hypothetical nanostructures are applied to materials processing and fabrication, which is a core skill underserved in modern materials engineering programs. Characterization of structure and properties, to develop both structure–property and process–structure relationships, is a core competency in itself, and nanostructures present challenges which even modern instruments cannot meet (see the Challenge of Characterizing Nanostructures reference). Finally, the turnkey process of PNPA ensures that technicians and engineers are exposed to all aspects of materials design, development, characterization, and support as a device moves from R&D into prototyping and manufacturing. PNPA develops "well-rounded" technicians.

PNPA defines the "four corners" of the engineering method, from applications/requirements to modeling and simulation of properties and material selection, to fabrication and characterization, to developing structure–property relationships and process–structure relationships. These four corners, not coincidentally, overlap the core functional groups in modern technology firms, from R&D (science and engineering) to early prototyping and device development (technology), to manufacturing and QA/QC. Characterization as a competency supports all four services, and includes process optimization, problem solving and failure analysis, and developing QA/QC systems for manufacturing support.

While education focuses on STEM (science, technology, engineering, and math) these four workgroups are better labeled as SETM, for Science and Engineering (R&D), Technology (prototyping), and Manufacturing. In the core areas of emerging industry, from energy to biomedical devices to advanced materials, all technology companies will own the process from R&D through manufacturing. Very few industries (Pharma may be an exception) can outsource manufacturing and remain competitive and benefit from downstream profit margins. The market size for PV panels, batteries, and advanced materials for automotive and aerospace applications are too large not to capture full cycle profits.

Industry surveys performed in both nanotechnology (2004) and clean technology (2008) by Foothill College, and nanotechnology (2002) by Dakota County Technical College (DCTC) found that inability to communicate and collaborate across functional workgroups (R&D, technology, manufacturing) was a key problem technology firms would like to solve. Much of the communication issues stem from lack of a common perspective, and a lack of understanding of the discipline of that perspective (especially science vs. engineering, R&D as compared to manufacturing). The PNPA rubric used as a focus across the nanomaterials program, and developed as a framework, and reinforced through internships, helps technicians practice a unified perspective on nanomaterials development. This ensures their productivity in R&D and manufacturing firms from day one, and additionally extends their ability to work in global collaboration environments (Boeing and other systems firms).

Global collaboration engineering and manufacturing firms are the new mantra, from aerospace to automotive to energy, clean technology, and even biomedical device applications. Today it is not uncommon to have R&D conducted in Silicon Valley, or Bangalore, or China, and have manufacturing located across three or more continents. The ability to use collaborative engineering tools, share documents and data across thousands of miles, and a dozen time zones, and in real time, is critical in achieving on-time product development and delivery and ensuring quality and support for complex products. The PNPA rubric, applied in a group setting (local or complex learning community), helps technicians prepare for the reality of working in teams, assuming key roles (based on competency), and sharing documents and data. Team-based learning is more fun and helps students work together to find solutions to problems which are not in textbooks, as is the case in most scenario-based curriculum.

PLO—Program Level Outcomes for Nanomaterials Engineering Program

The program learning outcomes (PLOs) for the nanomaterials engineering program are depicted in Figure 10.5. The goal of the nanomaterials engineering program is to provide materials scientists, engineers, and technicians with a framework to develop nanostructured materials with specific properties, an understanding of fabrication techniques, and competency using characterizations tools. Using scenario-based instruction, the practical use of fabrication and characterization tools is applied to a dozen nanomaterials, including diamond-like carbon, graphite and graphene, carbon nanotubes, CIGS thin-film materials, nanoparticles, dendrimers, self-assembled monolayers, biomedical materials (alloys and polymers), and cermets (electroceramics).

As students progress through the program they learn about key aspects of nanotechnology, including applications, structures, properties, fabrication, and characterization tools. In NANO51 students become familiar with the industry applications requiring materials with novel properties and nanostructured materials. In NANO52, students learn structure–property relationships, and specifically, how novel properties emerge from the unique position of specific atoms and interaction of physics in the extended electronic networks of those atoms. In NANO53, students learn the use of characterization tools to characterize the structure, composition and chemistry, and physical properties of materials. Additionally, analytical tools are used to monitor key process conditions to keep fabrication processes within specific

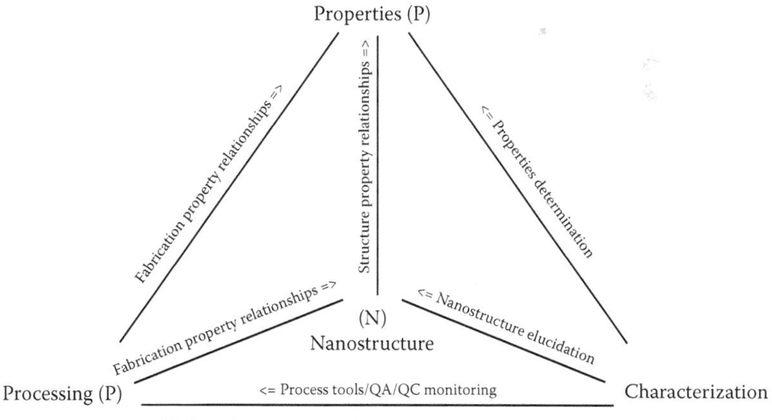

FIGURE 10.5
PLO

tolerances. Students learn methods of analysis as well as rote procedures for characterization, and specifically problem solving and experimental design. In NANO54, students will learn about three or more types of fabrication, such as thin film deposition, surface coating and modification, and alloying and annealing (sintering) to prepare nanostructured and high-performance materials. They will apply a range of process conditions to vary the observed appearance and properties of products and keep notebooks documenting experimental conditions and figure-of-merit properties.

The diagram is a directed graph—showing the relationships between and among fabrication, characterization, structure, and properties. What is important is the labeling of the edges connecting the vertices (nodes). These edges show the key learning outcomes students achieve in each course, and how they are combined in a holistic view of nanomaterials engineering. What our integrated program will achieve is the ability of technicians to create a framework by which they understand how fabrication produces a particular structural arrangement of atoms, and how the physics of this structural arrangement leads to specific electrical, mechanical, optical, and thermal properties.

Characterization tools inform technicians about how fabrication parameters cause a particular structural arrangement to occur, and how variations in process parameters produce predictable changes in structure. Using characterization tools to measure physical properties, scientists are then able to elucidate structure–property relationships. Combining the ability to "engineer" specific atomic, molecular, chemical, structural, and electronic networks, with knowledge of how properties emerge from structure, facilitates "engineering" of tailor-made properties (within physical limits). In future enhancements of our program, software tools will be used to help students (as well as incumbent technicians) apply structural modeling of materials to determine likely properties of nanostructures. Nanoexa, a computational nanotechnology developer, has developed a commercial tool for this effort.

In interviews with industry experts, the model we have shown in this diagram is a desirable framework for employees of start-up firms and practitioners developing nanostructured materials with novel properties. Our technicians will be trained to integrate this framework into all their learning and understanding of materials science. Deliberate application of this framework to each of the nanostructures below will also help technicians understand the importance of learning key material science concepts, including phase diagrams, diffusion, atomic, molecular, electronic, and crystalline structure, and the basic tenants of magnetism, thermal, mechanical, and electronic properties. Through exercises and demonstrations, students will consistently apply the integrated materials engineering framework to nanostructures (and nanosystems), reinforcing undergraduate materials science concepts.

Key Nanostructures and Nanosystems

- Diamond-like carbon (DLC)
- Graphite and graphene
- Carbon nanotubes
- Sputtered thin films
- CIGS thin-film material
- Self-assembled monolayers (SAMs)
- Plasma-deposited materials
- Surface-treated materials
- Nanoparticles
- Electroceramics

The PNPA rubric is central to the new nanomaterials pedagogy because it models the engineering process. Real-world engineering starts with developing an engineering and materials requirements document, specifying the materials and engineering properties needed for the application. From the requirements document, a succession of modeling and simulations are undertaken to determine what materials and nanostructures might provide the needed properties. In many, if not most cases, a structure or material with the exact required properties does not exist, but a closely related structure, or even a hypothetical structure, might provide a close enough match to start the engineering process.

In the next step, engineers develop a methodology to fabricate a small prototype sample, enough to perform some coarse physical properties measurements, which serve to validate structure and basic fabrication approach, and determine both process–structure and structure–property relationships.

Course Descriptions

Nanotechnology Survey Course (NANO51)

This is the first of four courses in the nanoscience and nanoengineering specialization, and introduces students to the "big picture concepts" of nanotechnology (SRI 2006/NCLT 2007), including an understanding of the physical forces at the dimension of an atom, quantum mechanics and electronic properties of atoms and molecules (and small systems comprised of atoms and molecules), chemical bonds and all types of bonding (covalent, ionic, metallic), fundamental nanostructures, basic fabrication techniques, characterization tools, key applications (industries), and societal impacts and grand challenge problems, including energy, water, food, medicine, computation, and transportation. A segment of this course will focus on careers in nanotechnology, including opportunities for technicians, incumbent workers adding new skills, and transitional and displaced workers with a science and engineering foundation looking to reenter the local high-tech workforce. The course uses a number of supplemental open educational resources (OERs) and electronic textbooks, including the Wikibook "Nanotechnology," which contains hundreds of individual articles on topics in nanotechnology. Additionally, we currently utilize, and will contribute to, nanoengineering pedagogy in the Wikiversity course "Nanotechnology." Our survey course has been in use for almost three years, but will be reworked using the integrated SLOs and PNPA rubric, especially to introduce the Application (A) component of PNPA. This course has also been approachable to high school honors students.

Students in NANO51 are currently introduced to hands-on use of the AFM/SEM at Foothill College, using the online course at AFM University,[21] and will use supplemental online training material in SEM (Hitachi) as we acquire and install a tabletop SEM as part of this project. Lab experiments in NANO51 now include preparing self-assembled monolayers (SAMs), using the Asemblon Educational kit, and reviewing XPS data collected from Stanford and EAG Labs. Students will also tour Stanford's Nanofabrication Facility (SNF) and Nanocharacterization Laboratory (SNL) as well as Evans Analytical Group (EAG Labs) in Sunnyvale, California. Additional tours of a carbon nanotube (CNT) fabrication facility (Unidym) in Menlo Park, thin film deposition facility (Southwall Technologies), in Palo Alto, and solar fabrication facility (Miasolé), in San Jose, California, will also be scheduled. Each of these firms and organizations are members of our formal and extended nanotechnology program advisory board.

A special emphasis of this course is an understanding of broader application areas, such as energy, water, bioengineering, nanomedicine, and advanced materials, the materials challenges in each of these fields, and how development of nanostructures is central to solving key problems in these fields. For example, if the cost and efficiency of photovoltaic panels were to improve by an order of magnitude, there would be an opportunity to displace coal-fired electricity without subsidies. This drives thin-film materials made of copper-indium-gallium-selenide (CIGS). What are the best nanostructures for CIGS, and how does thin film deposition by evaporative approaches differ from spin coat in manufacturing these materials at scale? These questions are an application of the PNPA rubric (process ⇒ structure ⇒ properties ⇒ applications) developed by NCLT and used by SRI and Foothill College in developing scenario-based curriculum to reinforce big picture concepts with industry applications.

After completing the nanotechnology survey course, students will be ready to proceed to the nanomaterials/nanostructures (NANO52) and materials characterization courses (NANO53), which precede the nanofabrication course (NANO54). Using industry-driven application scenarios, students understand the need for material innovation to create the desired performance properties

(P), and the applications-driven focus (A) to develop, characterize, and scale up fabrication (P) of new nanomaterials and devices. The PNPA rubric and scenario-based curriculum are central to course and program design.

Nanostructures and Nanomaterials Course (NANO52)

This course takes an in-depth look at nanomaterials discussed in Introduction to Nanotechnology, including the chemistry and physics of nanomaterials. The course will also focus on major classes of materials and nanomaterials, including glasses and ceramics, polymers and composites, metals and alloys, and other unique nanomaterials and structures, including carbon nanotubes, quantum dots, and dendrimers. Applications of nanomaterials, nanostructures, and finished devices to key technology areas of interest, especially the broader societal applications of energy, water, food, and medicine, are topical areas and part of the PNPA rubric.

This course serves two purposes within the engineering degree and nanoscience program. First, to introduce students to fundamental materials science concepts within the engineering paradigm of process ⇒ structure ⇒ properties. Second, to introduce students to the key nanomaterials and nanostructures; carbon nanotubes, nanoparticles, nanoclays, nanowires, dendrimers, and unique molecular complexes, built through DNA scaffolding, and molecular self-assembly processes (SAMs).

The first third of the course focuses on materials and materials science, including traditional materials, typical processing (heat treatments), and characterization of physical properties, (e.g., strength, elasticity, stress-strain curves, and electrical and thermal properties). Optical and scanning electron microscopes are employed in learning metallography and characterization of failure surfaces. The second third of the course builds on material fundamentals, developing concepts and skills in manipulating material structures, such as grain boundaries, fibers, phase, crystalline structure, crystallinity/amorphous character, etc. Characterization tools including AFM, SEM, AES, and XRD are used to introduce students to higher resolution techniques for understanding combined information for composition, chemistry, phase, and orientation. In the final third of the course, students will delve into nanomaterials, experimenting with nanoparticles, carbon nanotubes, dendrimers, and self-assembled monolayers (SAMs). Data from additional characterization tools will be introduced, including FTIR, XPS, HPLC, and NMR. These latter structures and instrumental techniques will be revisited in more detail in the fabrication and characterization courses, where more exacting measurements will be made.

Each new structure (N) is introduced within the context of a material, device, and or application (A), such that students are aware of both where and why a materials development effort is taking place. For instance, Nitinol™ and CoCr alloys are used extensively in biomedical stents for improving patient outcomes, and surface chemistry is key to effective drug delivery and healing. Carbon nanofibers are used in electronics, lighting and displays, and structural applications (nanocomposites), and optimizing carbon deposition and fiber surface chemistry are a continual challenge, especially when scaling up from R&D to manufacturing. Students will not only practice the PNPA rubric in learning about these materials, they will reinforce the process ⇒ structure ⇒ properties engineering paradigm within each learning unit. Additionally, they begin to understand the challenges of emerging from R&D into prototyping and early manufacturing scale up, and the importance of having baseline spectroscopy, composition and chemistry, and other structural data for the new and unique materials they are developing. Students also use electronic notebooks for recording, sharing, and submitting experimental laboratory reports.

Materials Characterization Course (NANO53)

The characterization course is where scenario-based curriculum is strongest. In this course we introduce the complete suite of characterization and process tools used by materials developers and materials analysts, providing invaluable insights into process ⇒ structure ⇒ property relationships in material and process development, as well as answers to problem solving questions in materials development, and failure analysis of components and devices. A strong instrumental and characterization background is the foundation of the most valuable technicians, who support R&D experiments and troubleshoot process engineering and manufacturing problems. Learning characterization tools in the context of real problem solving develops problem solving skills faster, as well as supporting deep learning of instrumental techniques and how to think through problems at all lengths of the innovation process. Industry-driven problems also help form a "foundation of practice" used by technicians in conducting materials characterization and failure analysis in industry laboratories. Our scenario-based curriculum reinforces the three dimensions of competency found in industrial labs (EAG Labs), which are the individual instrument skills, knowledge of the device and industry, and experience by approach (i.e., materials characterization, problem solving, and failure analysis). The problem found in many industries (Boeing/SRI) is that technicians are trained to run just one or two instruments, and typically look at just a few types of materials, and their span and depth of knowledge are limited by those techniques and samples. Our "holistic" approach is to introduce image, surface, organic, and structural analysis tools in the context of materials development, process troubleshooting, and failure analysis. By proximity to Silicon Valley, our "problem space" includes semiconductor, magnetic storage, biomedical device, energy, and thin films. These five industries all rely on advanced materials and nanostructures to stay on an ever-accelerating innovation curve.

Image analysis techniques combine optical and confocal microscopy, atomic force microscopy (AFM), and scanning electron microscopy (SEM) with metallography and image analysis in the nanostructures course. Typical materials problems will focus on metallography, grain size, particle size and distribution, adhesion/cohesion failure analysis, and characterizing brittle and ductile failure. Students practice AFM and SEM in a number of routine characterization exercises to develop QA/QC skills. Surface tools including XPS are principally used in characterizing surface composition and chemistry, especially for thin films and surface contamination, understanding the degree of bonding and orientation of self-assembled monolayers (SAMs), and correlating the effect of plasma surface treatments (nitrogen and oxygen plasmas) with changes in surface chemistry, wettability, and enhanced adhesion or biological growth and biocompatibility. Students observe AES analysis performed on semiconductors, grain boundary analysis, and use those data and measurements to evaluate oxidation and depth composition profiles of thin films. Additionally, students will combine image, surface, and structural techniques in practicing QA/QC protocols on electropolished 316L stainless steel, Nitinol™, and CoCr alloys. Combining AFM, SEM, XPS, and AES characterization of metals will introduce students to the synergies of multitechnique analysis, as well as help them develop their own protocols for analysis of metals. Organic analysis using FTIR, GC/MS, HPLC, and NMR will be included in a longer one-semester course offering, or as an optional or exchangeable unit for students with good organic chemistry skills. The characterization course uses electronic laboratory notebooks, LIMS (Laboratory Information Management Systems), Web-based shared image archives, and a combined class Wiki for disseminating information, all of which are standard practice in modern industrial laboratories.

Nanofabrication Course (NANO54)

In the final course of the series, students continue the PNPA rubric theme and take their overall knowledge of nanoscience and nanotechnology, exposure to real-world application, and materials development and build on their characterization prowess to add process to knowledge. Much of the work in this course will be conducted on small-scale simulation apparatus, such as vacuum deposition, solution chemistry, carbon furnaces, and fabless IC design for MEMS. Much of the project proposed here involves building out fabrication resources and curriculum, especially for thin film deposition, nanochemistry clean room apparatus, and software and computers for modeling and designing MEMS for fabrication in fabless ICs.

Thin film deposition is a core fabrication technology that is practiced across many industries; hence it will be a primary process focus in the PNPA rubric. Students learn about process tools associated with evaporative deposition, DC magnetron sputtering, chemical vapor deposition (CVD), plasma enhanced CVD (PECVD), metal organic-CVD (MOCVD), and spin coat technology. Students will practice simple metal deposition from a small tabletop bell jar DC magnetron sputtering system, characterizing the films using SEM,

TABLE 10.3

SLOs of the Integrated Nanoscience Program

NANO51: Introduction to Nanotechnology

- Scale and forces
  - Dominate forces at all scales of distance
- Emergence of properties at scale
  - Melting point, plasticity, thermal and electrical conductance
- Self assembly process
  - Crystals, molecular networks, biomolecules
- Atom as a building block of materials
  - Crystals, glasses, metals, liquid/networks
- Surface dominated behavior
  - Surface area vs. volume, surface properties vs. bulk, surface behavior and chemistry
- Role of quantum mechanics
  - Conduction, phonons, interaction with light
- Applications of nanotechnology/devices
  - Solar panels, fuel cells, semiconductors, ink
- Industries that use nanotechnology
  - Semiconductor, electronics, energy, medicine, advanced materials
- Characterization tools
  - Image (AFM/SEM), surface (AES/XPS), structure (XRD/TEM), and bulk (XRF/EDX/WDX)
- Societal issues — grand challenges
  - Energy, water, food, medicine, transportation

NANO52: Nanomaterials and Nanostructures

- Nanomaterials vs. "traditional" materials
- Nanostructures and novel properties
  - Nanofibers, nanoparticles
- Process → structure → properties → applications
  - Designing structures for end use properties
- Types of materials
  - Glass, ceramic, metal, alloy, polymers and composites,
- Types of properties
  - Strength, plasticity, thermal and electrical conductance, electromagnetic
- Fabrication basics
  - Fab facilities, tools, processes, CNT
- Processing
  - Heat treatment, quenching, alloys, composites, fibers
- Modeling and designing for desired properties
  - Computer modeling of structure
  - Computer modeling of properties
- Choice of materials and structures
  - Select for properties and applications
- Characterization tools for nanostructures/nanomaterials
  - Image, surface, composition, structural

NANO53: Nanocharacterization

- Instruments and characterization tools
  - (AFM/SEM), surface (AES/XPS), structure (XRD/TEM), and bulk (XRF/EDX/WDX)
- Types of analyses
  - Materials characterization, process development, failure analysis
- High vacuum and high voltage basics
  - Vacuum technology and safety, vacuum awareness, high voltage and safety
- Sample preparation and handling
  - Cleanliness, cleaning, dust and particles, vacuum considerations
- Instrument selection
  - Image, surface, composition, structure, or electronic/physical properties
- Data gathering, analysis, and tabulation
  - Instrumental techniques, data gathering and inspection, tabulation and interpretation
- Interpreting composition and chemistry, modeling structure
  - Building atomic and molecular structure from composition, chemistry, and x-ray data
- Using a LIMS, searching spectral databases
  - Using knowledge management tools to aid interpretation and future problem solving efforts
- Reporting data, writing formal industry reports
- Client management skills
  - Explaining results, gather experimental details, managing further experiments

NANO54: Nanofabrication

- Type of fabs
  - Silicon
    - MEMS
    - Wafers
  - Clean room basics
  - Air filtration, dust
- Safety basics
  - Vacuum equipment
  - High voltage
- Silicon fundamentals
- Deposition, masking, etching
- Virtual/physical tour of a silicon fab
- Fabless ICs
  - CAD, Microfabrica™ eFab
- MEMS basics
  - Silicon and polymer based MEMS
- Nanochemistry
  - Self-assembled monolayers
  - Dendrimers
  - Quantum dots
- Thin film deposition
  - Vacuum deposition
  - Sputtering, CVD/PECVD
  - Roll coating (web)
  - Spin coat
- Plasma deposition
  - Plasma equipment
  - Gas chemistry
- Surface modification
  - Chemical, gas, plasma

and later working as interns in larger commercial labs (solar technology). Students learn vacuum fundamentals and high voltage safety before working on commercial equipment. They tour Stanford's nanofabrication facility, as well as other commercial firms. Dendrimer synthesis will be performed in our organic synthesis laboratory, with FTIR, NMR, GPC, and HPLC characterization for purity and molecular weight determination. These advanced lab exercises reinforce both synthetic nanochemistry and analytical characterization methods, using nanomedicine as the application to tie together the PNPA rubric. Preparation of self-assembled monolayers (SAMs) is an additional example of nanochemistry, with coated materials used in oil recovery (interface sciences). Carbon nanotube (CNT) preparation reinforces gas handling, plasma chamber (plasma instrumentation), and prepares technicians to work around fiber synthesis gas, and in particular larger R&D scale-up and manufacturing facilities. These labs may be performed (observed) at partner institutions (e.g., Stanford University). Last, students are introduced to the design of MEMS using CAD and fabless IC approaches following the work of Sandia Labs and the University of New Mexico, which introduce students to semiconductor and silicon processing.

Table 10.3 delineates the expected SLO (Student Learning Outcomes) of the four integrated nanoscience programs.

Nano-Safety: The One Issue That Is Missing from the Education Equation

Walt Trybula, Ph.D., Contributor
IEEE Fellow and SPIE Fellow
Texas State University–San Marcos
Director, Trybula Foundation, Inc.

As mentioned previously in this chapter, we are creating a new type of worker. Addressing the challenges of nanotechnology is not trivial. A common saying in the early days of nanotechnology development was that if a person had a Ph.D. in physics, a Ph.D. in chemistry, and a Ph.D. in biology, he/she would be ready to start learning about nanotechnology. While the breadth of the developments covers these and more disciplines, the worker cannot be expected to be an expert in every field. However, the worker of the future in nanotechnology must be a knowledge worker. Industry-specific requirements are needed to guide the development of the education focus that needs to be provided to the students who are the future workers.

Over the years, I have been a leading proponent of addressing the issue of nanotechnology safety. The issue of safety is a critical issue, but one that is extremely difficult to quantify. An example will clarify the challenge.

According to the Environmental Protection Agency (EPA) guidelines, if someone breaks a low-energy light bulb, which contains micro amounts of mercury, the area needs to be cleared, protective clothing needs to be donned, and the debris carefully collected. One should never use a vacuum due to the exhaust being spread into the air. What is the procedure if one drops and breaks a container of nanomaterial? The answer that one would probably hear is that it depends. The situation is very serious because we do not know all the effects on people or the environment.

Consider the effect of three relatively common materials: silver, aluminum, and gold. Silver is a very interesting material in the nanoparticle form. Silver particles in the size range under 30 nm have some very beneficial properties. There is some indication that silver nanoparticles can prevent infection by destroying bacteria that could cause an infection. An advantage is that, unlike with antibiotics, the bacteria does not as easily develop an immunity to silver. This sounds great. However, if the silver nanoparticles are allowed to escape (be disposed of carelessly) into the environment, the particles will also kill "good" bacteria. This is a problem, and the EPA has provided some regulations in this instance. Aluminum nanoparticles are another good example of the "difference" in apparent properties in the nano realm. Aluminum is used for all types of structures. It is a lightweight metal with a reasonable strength. We have it all around us in our everyday lives. However, aluminum nanoparticles, below 30 nm, have a very interesting reaction when they are exposed to oxygen (as in air). The chemical expression is that there is a very energetic reaction. The lay person would say that it explodes! The last example is gold. Everyone has some exposure to gold. It has been used in civilization for millennia. Over time, experimenters have determined ways to process the metal. As we know today, the processing requires taking gold to its melting point, 1064.18°C, so that the metal can be purified. Something interesting happens to gold, as well as other metals, as the particle size gets smaller: The melting point changes! What we have known for millennia as constant actually changes. This change in melting temperature for gold starts to diminish below 20 nm and drops to below 500°C when the size approaches 1 nm. Why is this important? One reason is that a distribution of gold nanoparticles with a mean of 10 nm and a ½ width of ½ nm will behave differently from another distribution with a ½ width of 3 nm. The list of "unusual" occurrences of "known" material properties is very long, and most of the issues are still unknown.

This provides a challenge on how can we teach what we do not know. The next section in this chapter discusses how to address these unknowns. Working together with Dr. Dominck Fazarro since 2008, we have identified the elements required to provide a knowledge base for future nanotechnology workers. This effort has reached out to a number of two-year and four-year institutions to develop programs to address this critical need and then to be able to disseminate them throughout the country. Among the current efforts is one that involves both Texas State University–San Marcos (Drs.

Jitendra Tate and Trybula) and the University of Texas at Tyler (Fazarro) to develop nanotechnology safety insertion modules for existing courses. Another effort addressing nanotechnology safety modules for two-year programs involves UT Tyler (Fazarro) and Texas State (Trybula and Tate), Texas State Technical College, Austin Community College, Tulsa Community College, Dakota County Technical College, and the NSF Nano-link Center (Newberry). Another effort under development is addressing the needs of existing workers, and this involves Rice University (Dr. Kristen Kulinowski) and Texas State (Trybula).

The educational effort is only one part of a much larger effort. I have disseminated a "white" paper on NANO-SAFETY,[22] which was first released in 2007. There are four key sections in the white paper: (1) nanomaterial properties, (2) impact on people and the environment, (3) education of students and workers, and (4) business practices.

The issues involving nanomaterials include the fact that the complete knowledge of the nanoparticle properties and consequently their interaction with people and the environment is unknown and needs to be determined. The task of protecting people from the consequences of interaction with nanoparticles that could be harmful is challenging based on existing knowledge. Nevertheless, people and the environment need to be protected, a fact that requires an effort to evaluate what exists today, what is missing and needs to be developed, and how to initiate safety procedures based on today's knowledge. Improvements to the procedures will be made as new information becomes available. "NANO-SAFETY can be defined as creating an environment where both people and the environment are protected from exposure to harmful substances. NANO-SAFETY can be defined as creating a modus operandi where both people and the environment are protected from exposure to harmful substances."

With respect to all the significant effort being done on materials understanding and the nano-health issues, the most critical item is to develop and train workers in the methodology of the proper handling of nanotechnology. When the semiconductor was developing and the major manufacturers understood that they needed to have a means of handling the large mixture of very toxic chemicals, the large companies formed an organization to addresses the issues and provide for protection of the people and the environment. Nanotechnology has a more diverse group of users and does not have the size and the funds that the semiconductor industry did. This exposes a weakness in the development of nanotechnology efforts. Without large companies that can provide a leadership role in the development of the methodology of safely handling nanomaterials, the emerging industry is apt to create some difficult situations that have negative impact on people and/or the environment. This must not happen. Consequently, we have established educational efforts to develop training in the handling of nanomaterials.

Lateral Diffusion of NanoEducation: Developing the New Workforce

Dominick Fazarro, Ph.D., Contributor

As we move further in the twenty-first century, new technologies are emerging faster than ever. The emergence of nanotechnology may be considered the new "industrial revolution." As stated in the paper *Future Shock: What Would a Nanotechnology Curriculum Look Like?* (Fazarro and Kornegay, 2008):[23]

> We must prepare the 21st century workforce to compete globally with emerging technologies. In order to prepare for nanotechnology, educators must be proactive and not reactive to new technologies for developing cutting-edge programs at post-secondary schools as well as for workforce centers.

Dr. Mihail Roco, NSF Senior Advisor on Nanotechnology is a strong proponent of "nano workforce education." Roco stresses the training of people is critical for long-term success in the field of nanotechnology (Roco, 2001).[24] An estimate of 2.5 million nanoworkers will be needed globally in the fields of nanotechnology and nanoscience (Palma, 2007).[25]

There is a need for a comprehensive approach to bring awareness at all levels of education, according to Fonash (2001).[26]

The key to increasing awareness is informed school educators and counselors. They must appreciate the role nanotechnology will have on society, understand what is required to prepare the nanotechnology workforce, and successfully "get the word out" to students and parents. A lateral diffusion approach would be used to transfer knowledge from research to industry, to academia, including middle, secondary, and postsecondary levels. To achieve this innovative thinking requires the development of a pipeline of information that can be easily transferred to all levels of educational systems in the United States. Industries, universities, federal and state governments, and workforce centers must create a "multi-collaboration" agreement to develop "educational pipelines." However, there are challenges that must be met along with changing the paradigm of how we think about education as a whole. Meaning, we must come to terms with how we want to shape our future for the next 20 years. State, federal, and academic leaders tend to react from a complacent stance to do something based on a "ripple effect." Our current U.S. public schools are lagging in math and science, which may hinder the growth of a nano workforce (Fonash, 2001). Furthermore, to add to the lagging of math and science, administrators and teachers are not aware of the field of nano or have little interest because of the huge demand of testing and the inability to creatively add new knowledge. Leaders and stakeholders must become more "proactive" instead of "reactive" to move technology and develop people for a long-sustained and prosperous economy. Administrators at postsecondary institutions must be risk-takers to push innovative curriculum into the classrooms.

There are professionals and educators throughout the United States that are catalysts or "seed planters" to push the nano education and workforce movement forward. The major goal is to assist in a well-prepared nano workforce, which will be able to produce innovative products in medicine, green technology, automotive, and energy which can benefit society (Trybula, Fazarro, Kornegay, 2009).[27]

Dr. Walt Trybula, Texas State University, San Marcos, and I have been working on implementing nano programs at postsecondary schools in Texas. We have been presenting, collaborating, speaking with state officials, and writing grants to move the nano effort forward. The state of Texas is the ideal location to create the first truly nano workforce in the United States. With its diverse population, abundance of secondary and postsecondary schools and nano-based industries, creating the nano workforce will soon, in the near future, become reality. Examples of our efforts include implementing a nano-management focus in the Department of Human Resource Development and Technology at the University of Texas at Tyler. Dr. Trybula has been working with a number of faculty to ensure that we are providing the basis for educational experiences that are relevant to the workplace needs. He has also promoted the nanotechnology issues through various presentations, workshops, and other means of information dissemination including webinars. Dr. Trybula is currently working with Dr. Kristine Kulinowski, Director of the International Council of Nanotechnology, Rice University, on the creation of various "short" courses to provide the appropriate educational experience for those already in the workforce. We are also working on a lateral diffusion model to develop a "nano incubator" where researchers, educators, and workforce centers can meet to design nano curriculum. This model is based on mutual partnership with developing goals, objectives, and working roles to develop the nano workforce in Texas (see Figure 10.6).

The emergence of a nanotechnology workforce is rapidly building in other countries, which already identified the social and workforce implications. Community colleges in California, Minnesota, Oklahoma, Pennsylvania, Texas, and other states have implemented nanotechnology technician programs and are currently working with workforce centers and industry. However, we as educators and the industry sector must work on lateral diffusion of knowledge to all facets of society.

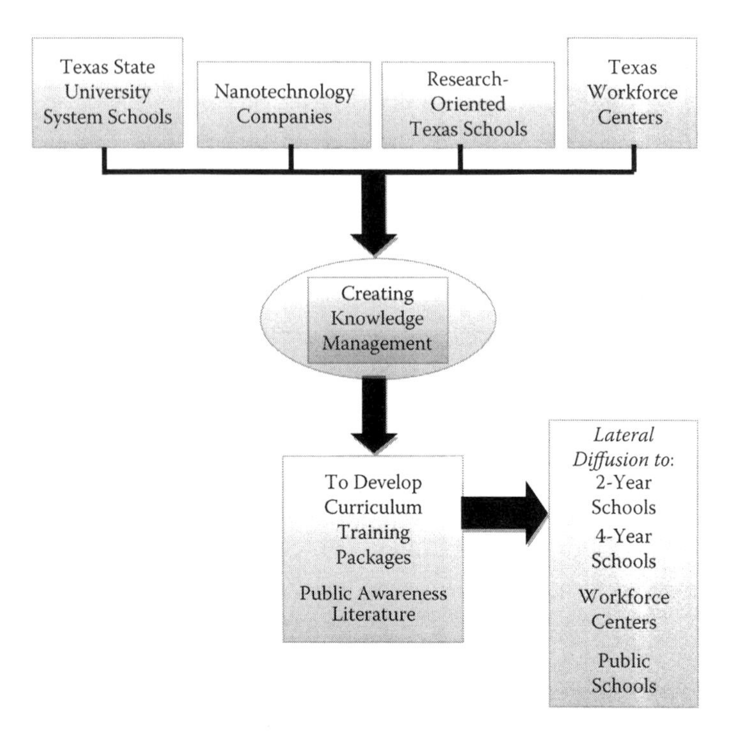

FIGURE 10.6
Texas State University System Nanotechnology Task Force collaboration model. From "The Evolution of Nanotechnology Education: Lateral diffusion for the 21st Century Workforce," by Dominick E. Fazarro, paper presented at the NanoTx Conference, Dallas TX, October 2–3, 2008, and by Trybula, W., Fazarro, D., and Kornegay, A. (2009). The emergence of nanotechnology: Establishing the new 21st century workforce. *The Online Journal for Workforce Education and Development*, 4(3), 1–10.

References*

1. NSF Workshop Report Partnership for Nanotechology Education NSF Award EEC 0805207, http://www.nsf.gov/crssprgm/nano/reports/nsfnnireports.jsp
2. http://www.cneu.psu.edu/nePublications.html#capstoneVideo
3. http://www.pct.edu/k12/
4. http://www.cneu.psu.edu/edOutreachSecEd.html#camps
5. http://www.pct.edu/degreesthatwork/nanotechnology.htm
6. http://www.cneu.psu.edu/edOutreachPD.html
7. http://www.cneu.psu.edu/edToolsActivities.html
8. http://www.cneu.psu.edu/edToolsVideos.html
9. http://www.nano4me.org/
10. http://nanotechworkforce.com/about/cluster_initiative.php
11. http://nanotechworkforce.com/curriculum/syllabus.php

* All links active as of August 2010.

12. http://nanotechworkforce.com/curriculum/modules.php
13. http://www.dctc.edu/future-students/programs/nanoscience-technology.cfm
14. http://www.nanoprofessor.net/
15. http://cnse.albany.edu/
16. http://cnse.albany.edu/research/profiles.html
17. http://cnse.albany.edu/Nano_for_Kids.html
18. http://www.foothill.edu/programs/programs.php?rec_id=564
19. *Nanoscience: A Vehicle for a Goals-Oriented Science Education Final Report,* July 2007, SRI Project Number 17699
20. http://www.nanoexa.com
21. http://www.afmuniversity.org
22. http://www.nano-safety.info
23. Fazarro, D. and Kornegay, A. (2008, December) Future Shock: What Would a Nanotechnology Curriculum Look Like?, Paper presented at the Association of Career and Technical Education (ACTE), Charlotte, NC.
24. Roco, M. (2001). International Strategy for Nanotechnology Research and Development, *J. of Nanoparticle Research,* Kluwer Acad. Publ. 3 (5–6), 353–360.
25. Palma, D. (2007). K–12 nanotechnology education outreach for workforce Development: The Georgia Institute of Technology Model. Retrieved from http://nanohub.org/resources/2251/supportingdocs.
26. Fonash, S.J. (2001). Education and training of the nanotechnology workforce. *Journal of Nanoparticle Research,* 3, 79–82.
27. Trybula, W., Fazarro, D., and Kornegay, A. (2009). The emergence of nanotechnology: Establishing the new 21st century workforce. *The Online Journal for Workforce Education and Development,* 4(3), 1–10. Retrieve from http://wed.siu.edu/Journal/VolIIInum4/Article_6.pdf/

11

Informal Science Resources

> The most exciting phrase to hear in science, the one that heralds new discoveries, is not "Eureka!", but "That's funny ..."
>
> **Isaac Asimov (1920–1992)**

The Nanoscale Informal Science Education Network (NISE Net) is a national community of researchers and informal science educators dedicated to fostering public awareness, engagement, and understanding of nanoscale science, engineering, and technology. Rapid advances are revolutionizing medicine, computing, materials science, energy production, and manufacturing. Yet, to the general public, these advances can be invisible or difficult to understand.

The NISE Network was created to engage the public in discussions concerning advances in nanoscale research, to develop programs and exhibits that capture the imagination and interest of young people, and to foster new partnerships among research institutions and informal science centers.

In 2005, the Museum of Science, Boston, the Exploratorium, and the Science Museum of Minnesota joined together with a dedicated group of partners and advisers to develop NISE Network. Funded by the National Science Foundation[1] through a $20 million, five-year cooperative agreement, the Network supports a variety of activities. Programs are designed as informal education to share professional resources including exhibits, programs, forums, media, and other tools to engage the public, and to provide the public with information and activities about nano, both online and in their neighborhood. The network is composed of eight Regional Hubs[2] listed on their Web site for your convenience.

A focal point of the work is NanoDays, coordinated as a week of community-based educational outreach events across the nation on a designated week each spring. Specifically designed to raise public awareness in grades K–12 of nanoscale science, technology, and engineering, the events have been very successful. In 2008, the NISE Network distributed 100 physical NanoDays kits to museums and university research centers throughout the United States. In 2009, 225 NanoDays kits were distributed to locations throughout the United States. Digital versions of kits[3] are available for download from the Web site as well.

The regional hubs facilitate partner interaction in the Network, help museum educators connect with researchers and each other, host regional workshops and meetings, and provide support to institutions in their region. To become involved with the Network, find your region on the map,[4] and contact your regional hub. The Web site also has a page of Introductory Materials[5] for K–12 teachers.

A Catalog of Programs[6]

"Programs" are public interactions facilitated in person by museum professionals. Here you will find all the materials you need to host many kinds of programs, including large lectures, small floor demonstrations, comedy and theater pieces, and quiz and game shows.

Cornell University Informal Outreach

Nanooze Magazine and Interactive Learning Site for K–12[7]

Nanooze is a magazine created to get kids excited about science and nanotechnology by revealing to them that the world that is too small to see is full of interesting stuff. It shows kids that scientists and engineers are beginning to understand this world and learning how to change things at the nanoscale level because, as the Web site says, "cool things happen when you make things really, really small. Like gravity hardly works, and there is no color!"

The magazine covers more than just nanotechnology, however, and describes many different things that are happening in other fields of science. Students, teachers, and parents can download the magazine, play the NanoCure game online, visit the blog, or go to the scientist question-and-answer section. Every week there are new articles and reports on the latest developments in science, and stories about real scientists and what they do.

Nanooze—now available in English, Spanish, Portuguese, and Swahili—is supported by the National Science Foundation through the National Nanoscale Infrastructure Network and especially the Cornell Nanoscale Science and Technology Facility.

Rensselaer Polytechnic Institute Presents the Molecularium™

Teachers' Discovery Guide and Teachers' Resource Guide Grades 5–8[8]

Riding Snowflakes is a state-of-the-art, computer-generated planetarium show designed to spark young children's interest in the atoms and molecules that constitute our world. Supported by a grant from the National Science Foundation (NSF), the show integrates advanced scientific simulations into an immersive educational animation. The Molecularium is part of the educational and outreach program of Rensselaer's NSF-funded

Nanoscale Science and Engineering Center (NSEC) for Directed Assembly of Nanostructures.

PBS DragonflyTV

DragonflyTV is an Emmy® Award-winning multimedia science education program combining television, community outreach, the Internet, and fun. Produced by Twin Cities Public Television (TPT), creator of the family science series Newton's Apple, it engages children in accessible, hands-on science activities. DragonflyTV is broadcast nationwide on PBS stations presenting real-life science investigations featuring children who communicate the infectious excitement of their discoveries.

DragonflyTV extends beyond television, offering a variety of standards-based learning tools. The companion interactive Web site provides an annual children's science magazine, educators' guides, and outreach materials available to schools, homes, and community organizations.

Education Standards—Inquiry-Based Grades 4–6

The philosophy of DragonflyTV is rooted in a strong commitment to inquiry-based learning, which is at the heart of current reforms in science education. Incorporating the National Science Education Standards, the programming communicates basic science concepts appropriate for grades 4–6, promoting diversity and the idea that all children are capable of understanding and "doing" science.

Resources[9]

The activities page has riddles and a Nanobot game, along with a section of "DO IT" experiments for kids to try at home.

The Lawrence Hall of Science—University of California, Berkeley

The university hosts two nanotechnology-focused Web sites with activities, which were funded as separate projects by the National Science Foundation (NSF), for students in grades K–12.

Nanozone[10]

The Nanozone is designed for students ages 8 though 14 and explores the basics of nanotechnology with interactive games, videos, and comics. The site also features interviews with scientists and teacher classroom materials for lessons on nanotechnology.

Nanotechnology: The Power of Small[11]

Developed for high school students and adults. The Power of Small Web site focuses on the ethical questions surrounding the rapidly advancing field of nanotechnology. The site features videos from a three-part Fred Friendly panel, where experts debate real-life scenarios, and an online forum to share ideas and discuss hypothetical situations involving the application of nanotechnology. View the outcome of *Nanofutures,* a series of deliberative democracy forums, visit a range of nanotechnology-related events at Science Centers around the country, or just gather information about this powerful new suite of technologies and the opportunities and challenges they present.

Lawrence Berkeley Labs—Nano*High Program[12]

The University of California's Lawrence Berkeley National Lab is now offering high school teachers and students—with or without a science background—a series of free Saturday morning lectures and laboratory tours. Students and teachers will meet and talk with world-renowned UC Professors and Berkeley Lab scientists and graduate students, learning "about their research into the world of the ultra small and how it will affect our future." Nano*High talks are aimed at all high school students, from those already committed to careers in science to those committed to poetry, history, philosophy, or to figuring out what they want to be committed to.

Nano*High Lectures[13] are placed on You Tube and discuss many topics of interest to the public.

ChemSense—Visualizing Chemistry[14]

ChemSense studied students' understanding of chemistry and developed software and curriculum to help students investigate chemical systems and express ideas in animated chemical notation. The ChemSense software is now available for download, and the source code is available as open source.

Activities and Use

The ChemSense curricular framework highlights collaborative investigations, representational competence, and chemical change. ChemSense curricular modules are codeveloped by teams of researchers, developers, and teachers. Each module contains learning objectives, hands-on experiments, and integrated chemistry tools. See examples of student and teacher work and sample curriculum activities.

Key Chemical Themes: Chemical Change

Curriculum is designed around a set of five key time-dependent dimensions that are identified as associated with the particulate nature of matter and chemical reactions: change in (a) connectivity, (b) molecular geometry, (c) aggregation, (d) state, and (e) concentration. Taken together, these dimensions begin to portray the molecular world imagined by chemists to account for observable phenomena. All involve changes in molecular and supramolecular structure that correspond to critical aspects of chemical reactivity. In addition, these time-dependent dimensions cut across more traditional chemical topics, such as acid–base reaction, electrochemistry, solubility, kinetics, and thermodynamics.

Sample curriculum—38 activities[15] are available.

NanoSense[16]—The Basic Sense behind Nanoscience

The NanoSense project (2004–2008) addressed the question of how to teach nanoscale science at the high school level. Working closely with scientists and educators, the project created, tested, and disseminated four curriculum units to help high school teachers and students understand nanoscale science. The project hosted workshops to introduce teachers to the materials and held working meetings with experts and practitioners to identify and clarify major concepts and learning goals for nanoscience education.

NanoSense materials were developed by SRI International (formerly Stanford Research Institute) with support from the National Science Foundation. Drawing on concepts from physics, chemistry, and biology, NanoSense units reflect the interdisciplinary nature of nanoscience, emphasize fundamental nanoscience concepts including size and scale and surface dominance of reactions, and explore applications of nanoscience and how they could affect society and policy.

Teaching Resources

Each unit[17] includes background materials and directions for teachers, activities and instructional materials for students, and embedded assessments. Available units include:

- Size Matters: Introduction to Nanoscience
- Clear Sunscreen: How Light Interacts with Matter
- Clean Energy: Converting Light into Electricity
- Fine Filters: Filtering Solutions for Clean Water
- Alignments with Curriculum Topics

The alignments with curriculum topics in chemistry, physics, biology, and environmental science are also available for download and provide mappings between NanoSense materials and standard science curriculum textbooks.

"When Things Get Small"[18]—UCSD TV

This 27-minute program teaches viewers about nanoscience—technology at one-millionth of a millimeter—through an entertaining mix of science and humor. Produced for University of California Television (UCTV) by Not Too Serious Labs, it departs from the typical science-for-television fare by using illustrative concepts that include a stadium-sized bowl of peanuts, a magic tennis ball, and shrinking elephants to describe the quest to create the world's smallest magnet. The program is now available in Spanish and Portuguese and can also be seen on YouTube. UCSD-TV also has more documentaries and programs available on nanoscience and physics.

Understanding Science—University of California, Berkeley Museum of Paleontology

The Understanding Science site was produced by the UC Museum of Paleontology of the University of California at Berkeley, in collaboration with a diverse group of scientists and teachers, funded by the National Science Foundation. Understanding Science was initially inspired by their work on the Understanding Evolution project, which highlighted the many misconceptions regarding evolution. Most spring from misunderstandings of the nature of science. Research indicates that students and teachers at all grade levels have inadequate understandings of the nature and process of science, taught as a simple, linear, and nongenerative process. The public needs to assess conflicting representations of scientific evidence in the media. Understanding Science takes an important step toward meeting these needs.

The immediate goals are to (1) improve teacher understanding of the nature of the scientific enterprise, (2) provide resources and strategies that encourage and enable K–16 teachers to reinforce the nature of science throughout their science teaching, and (3) provide a clear and informative reference for students and the general public that accurately portrays the scientific endeavor.

Teacher Resources Categorized[19]

Grade levels: K–2, 3–5, 6–8, 9–12, 13–16.

Students should come away from our classrooms with an appreciation of the natural world—fascinated by its intricacies and excited to learn more. They should view and value science as a multifaceted, flexible process for better understanding that world. Such views encourage lifelong learning and foster critical thinking about everyday problems students face in their lives. Teachers can cultivate these ways of thinking in their students through science instruction that accurately and enthusiastically communicates the true nature of science and encourages students to question how we know what we know.

Fortunately, fostering such understandings need not require reorganizing the entire curriculum. Simple shifts in how content and activities are approached can make a big difference in overcoming student misconceptions and building more accurate views of the process of science. Educational research supports the following strategies for teaching about the scientific endeavor.

- Make it explicit: Key concepts regarding the nature and process of science should be explicitly and independently emphasized. Engaging in inquiry and studying the history of science are most helpful when the nature-of-science concepts they exemplify are explicitly drawn out in discussion and interactions.
- Help them reflect: Throughout instruction, students should be encouraged to examine, test, and revise their ideas about what science is and how it works.
- Give it context, again and again: Key concepts about the nature and process of science should be revisited in multiple contexts throughout the school year, allowing students to see how they apply to real-world situations.

Resource Library[20]

The University at California at Berkeley has assembled a variety of resources to help put these strategies into action in classrooms. To get started, explore the resource library.

NISE Network Videos, Audio, and Podcasts[21]

The Twinkie Guide to Nanotechnology

Andrew Maynard of the Project on Emerging Nanotechnologies provides an introductory guide to nanotechnology.

NanoNerds

Variety of fun, educational videos about nanotechnology produced by the Museum of Science, Boston.

Earth & Sky

Variety of 90-second radio mini-documentaries on nano for distribution to radio stations nationwide, and Earth & Sky Clear Voices for Science nano episodes.

Sound Science

The Oregon Museum of Science and Industry's podcast answers questions about nanotechnology, including: what is it, what can it be used for, can it build a space elevator, and what is gray goo (NanoDays episode, 4/4/08).

Small Talk Podcasts

SmallTalk is a podcast series chatting about nanotechnology with leading scientists, thinkers, artists, writers, and visionaries produced by the Exploratorium.

NPR's Science Friday

Host Ira Flatow focuses on various nano topics and materials science topics: Nobel Chemist Harry Kroto and Buckyballs (March 20, 2009), Nanotechnology (June 15, 2007), Nanoantenna Sheets Harvest Energy (August 22, 2008), Nanotech: Small Things Considered (August 11, 2000).

The Exploratorium[22]—San Francisco, CA

Housed within the walls of San Francisco's landmark Palace of Fine Arts, the Exploratorium is a collage of hundreds of interactive exhibits in the areas of science, art, and human perception.

The Exploratorium stands in the vanguard of the movement of the "museum as educational center." It provides access to, and information about, science, nature, art, and technology.

This unique museum was founded in 1969 by noted physicist and educator Dr. Frank Oppenheimer, who devoted his efforts to it—and was its director—until his death in 1985. From 1991 until 2005, the museum was led by renowned French scientist and educator Dr. Goéry Delacôte. In May 2006, nationally known science education and policy expert Dr. Dennis M. Bartels was named Executive Director.

The Exploratorium educational interactive Web site has been online since 1993 and was one of the first science museums to build a site on the Internet. It now contains over 18,000 award-winning Internet pages exploring hundreds of different topics, serving 20 million visitors a year.

Microscopic Imaging Station[23]

The Microscopic Imaging Station is supported by a Science Education Partnership Award (SEPA) from the National Center for Research Resources, National Institutes of Health, and the David and Lucile Packard Foundation. Students can view blood cells, sea urchins, stem cells, zebra fish, neurons, cancer cells, frogs, and planaria in full color with an educational story for each image. A download file of each image is provided for teachers' use in the classrooms.

Activities[24]

Classroom Explorations: Introduces students to unique life science activities that let them work with our research-quality microscopic images and videos.

Science Museum of Minnesota (SMM)

Explore—Online Learning Activities [25]

SMILE Pathway features all the best activities for students listed on one page. SMILE is an online collection of math and science activities available to anyone, free of charge. Whether your "classroom" is an active volcano, the shark tank at the local aquarium, or your own kitchen table, you have come to the right place.

Science Buzz[26]

A blog on nanotechnology developed by the University of Minnesota as a partner of the NISE Network of museums.

Learn at Homeschool Classes[27]

The Science Museum of Minnesota has exciting homeschool classes available for families looking to supplement their science curriculum, explore physics, biology, engineering, and more with our specialized equipment and expert teachers, and network with other homeschooling families.

Science and Engineering Classes

The Science Museum of Minnesota offers in-depth, multiweek classes in a variety of science disciplines. These classes are paired with an engineering class that links to the content area. They also offer the acclaimed Engineering is Elementary curricula from the Museum of Science in Boston. This program fosters engineering and technological literacy among children. The curricula integrate engineering and technology concepts and skills with elementary science topics, literacy, and social studies. It features fun, in-depth activities that introduce the design process and get kids doing three-dimensional problem solving.

Boston Museum of Science—Partnership with NSEC Harvard

NSEC Informal Education and Public Engagement

The participants and their colleagues develop innovative science communication strategies for enhancing public understanding of research in nanoscale science and engineering, engaging a broad range of audiences at the Museum of Science and elsewhere.

Monthly Live NECN Cablecasts

"Sci-Tech in the News" features monthly nanotechnology stories cablecast live from the Museum's Current Science and Technology stage via New England Cable News, reaching up to 2.8 million homes and businesses.

Guest Researcher Appearances: Up Close and Personal

Not only are museum audiences attracted to the series of weekend encounters with real NSEC researchers, the researchers themselves seem to enjoy the opportunity. Guest researchers delight more than 400 museum visitors annually with tales from the nano laboratory bench.

Talking Nano (6 DVDs)[28]

Talking Nano was filmed at and produced by the Museum of Science, Boston, in association with the NSF Center for High-rate Nanomanufacturing, the Harvard NSF Nanoscale Science and Engineering Center, the Wilson Center Project on Emerging Nanotechnologies, and the NSF Nanoscale Informal Science Education Network (ESI-0532536).

Resources for Teachers[29]

The museum has a series of live presentations, podcasts, and videos available on the site.

Rice University—Center for Biological and Environmental Nanotechnology (CBEN)

Informal Community Outreach Programs[30]

The Matter Factory and Children's Museum of Houston (CMH)

CBEN partnered with the Children's Museum of Houston (CMH) in development of The Matter Factory as part of the museum's expansion. CBEN worked with CMH design staff to provide nanotechnology and materials-science expertise to guide planning, construction, and evaluation, especially on a carbon self-assembly exhibit.

The exhibit is designed to expand the age range of students attending the museum with displays that target 8- to 12-year-old students. The main components of the exhibit are:

- **Sensory entry:** Simple hands-on and body-on activities introduce properties of matter.
- **Object sorting:** A group activity area that teaches basic materials properties to younger visitors.
- **Property testing:** Activity area introduces older children (8- to 12-year-old) to advanced properties and states of matter.
- **Smart materials:** Activities demonstrate new materials and new applications.
- **Nanoscience:** A "scale-shift" area where children explore matter at the micro- and nano-level.

Nano Days

CBEN graduate students, faculty, and staff worked with the CMH staff to develop and host the NISE-Net Nano-Days program on March 28, 29, and 30, 2008. CBEN organized and participated in the "Ask the Scientist" sessions, in a discussion-based format designed to educate the museum attendees about nanotechnology and its relevance. Graduate students entertained and taught students about applications of nanotechnology and showed how their research might impact students' lives. In addition, graduate students hosted demonstrations and activity tables that included topics such as Seeing Scale, Smelling Scale, Invisible Sunblock, Environmental Cleanup, and made a three-story balloon sculpture of a nanotube. Information about plans for future Nano Days participation was not posted on the Web site.

Check the Community Programs page for future events.

What Is Cool Science?[31]

At Cool Science, scientists answer questions of all kinds (Ask a Scientist). Young scientists are encouraged to get their hands dirty ... virtually (Curious Kids). High school and college students are shown new approaches to cutting-edge science topics (BioInteractive). Educators are provided with a host of innovative resources they can use in their classrooms (For Educators). Explanations are provided of what it takes to become a scientist (Becoming a Scientist). An undergraduate science discovery project is showcased, that may one day change the way science is taught (SEA). Explore the many cool features of Cool Science for teachers and students.

Resources are provided as an outreach program of Howard Hughes Medical Institute.

Science News for Kids[32]

The Science News for Kids Web site, funded by grants from several corporations and foundations, enhances the usefulness of Science News in the middle-school classroom and offers recreational reading and activities for students interested in science. A section of the Science News for Kids Web site is devoted to particular interests. At present, there were six such zones: a weekly brainteaser for those who enjoy solving and inventing puzzles (PuzzleZone), entertaining science-fiction composition exercises for those interested in writing (SciFiZone), and weekly science fair profiles along with science project ideas and tips (ScienceFairZone). The GameZone contains a small selection of logic and memory games, implemented as Java applets. The TeacherZone has materials, including question sheets related to the feature article of the week, so that teachers can bring science news topics to the classroom. The LabZone features a weekly hands-on activity or science project idea.

UnderstandingNano Web Site Offers Lesson Plans for Educators

The UnderstandingNano Web site has posted five new nanotechnology lesson plans[33] in time for the 2010–2011 school year. The site now offers lesson plans for both high school and middle school grade levels on three topics: Introduction to Nanotechnology, Nanotechnology in Medicine, and Environmental Nanotechnology. Each lesson plan includes a corresponding student handout and all plans are available for anybody to use.

References*

1. http://www.nano.gov/html/edu/home_edu.html
2. http://www.nisenet.org/about
3. http://www.nisenet.org/nanodays/kit/digital
4. http://www.nisenet.org/community
5. http://www.nisenet.org/community/K–12 -teachers
6. http://www.nisenet.org/catalog/programs
7. http://www.nanooze.org/
8. http://www.moleculestothemax.com/Educators.html
9. http://www.pbs.org/parents/dragonflytv/activities.html
10. http://www.nanozone.org
11. http://www.powerofsmall.org/
12. http://www.lbl.gov/msd/nanohigh/index.html
13. http://www.lbl.gov/msd/highlights/youtube-msd.html?p=954C311769FAF77E
14. http://chemsense.org/
15. http://chemsense.org/classroom/activities.html
16. http://www.nanosense.org/
17. http://www.ucsd.tv/getsmall/
18. http://www.nanosense.org/activities/html
19. http://undsci.berkeley.edu/teaching/index.php
20. http://undsci.berkeley.edu/resourcelibrary.php
21. http://www.nisenet.org/public
22. http://www.exploratorium.edu/
23. http://www.exploratorium.edu/imaging_station/index.php
24. http://www.exploratorium.edu/imaging_station/activities.php

* All links active as of August 2010.

25. http://www.smm.org/explore/
26. http://www.sciencebuzz.org/buzz_tags/nanotechnology
27. http://www.smm.org/homeschool/
28. http://talkingnano.net/
29. http://www.mos.org/topics/nanotech_and_nanomedicine
30. http://cben.rice.edu/communityprograms.aspx
31. http://www.hhmi.org/coolscience/
32. http://www.sciencenewsforkids.org/
33. http://understandingnano.com/nanotechnology-lesson-plan.html

12

Overviews: Global Nanotechnology Initiatives and Resources

> We are at the very beginning of time for the human race. It is not unreasonable that we grapple with problems. But there are tens of thousands of years in the future. Our responsibility is to do what we can, learn what we can, improve the solutions, and pass them on.
>
> **Richard Feynman**

Growth of Nanotechnology Education and Initiatives Globally

The first Nanotechnology Conference addressing *Human Resources Development in Nanotechnology*[1] was held in 2003. It was hosted by Dr. Joydeep Dutta of the Asian Institute of Technology (AIT) in Thailand, partnering with Dr. Jurgen Schulte, of the Asian Pacific Nano Forum (APNF), with support from the Microelectronics School of Advanced Technologies (MTEC), the Thai Academy of Science and Technology (TAST), and the National Science and Technology Development Agency (NSTDA).

The goals of the conference were to begin exploring and anticipating future educational and training needs in the emerging field of nanotechnology for the Asian Pacific region. Specific goals were set to identify and discuss education requirements, specific programs, and development and implementation of curriculum; also to identify curriculum for present and future industry workforce needs in their region and look at e-education possibilities, and to bridge the gap amongst available academic expertise in different countries.

To address these areas of inquiry, many models were presented by professors from Sweden, Switzerland, Australia, Vietnam, Malaysia, New Zealand, Thailand, and Japan. The speakers broke into working groups to foster discussions regarding the necessity of cross-cultural communication, not only between countries, but within the academic community. Everyone felt it was necessary to integrate chemistry, physics, biology, and engineering to understand the nanoscale of science. Development of completely new courses was also discussed for long-term solutions, while module expansion of existing curriculum was decidedly sufficient for the near term.

Dr. Dutta[1] took the lead and, immediately following the Conference, developed the first free e-learning Web site for students globally. In 2004, he set up doctoral research, and the AIT became a Center of Excellence in Nanotechnology in 2006. During this time he coauthored textbooks and launched the master's degree program in 2009. Dr. Dutta called his efforts "poor-man's nanotechnology," making use of inexpensive wet-chemical methods to fabricate innovative materials and futuristic device components.

The conference brought to the forefront an understanding that each country in the region must find their niche products and expand the research in that area of expertise. Thailand understood this premise, and they continue the philosophy to this day. The Thailand National Nanotechnology Center uses the phrase "to create niche products, and processes with Nanotechnology" as their byline. A recent article[2] describes their success and the next phase of their plans for 2010.

> A national strategic plan for nanotechnology, launched in 2007, had been a success initially in the textile, chemical and medical sectors. Among new products are fabrics and Thai herbal medicines that are both "nano-coated." Thailand is expanding its nanotechnology strategy into the energy and agriculture sectors after reporting success in the first phase of its national nanotechnology policy. Developments such as nano-based solar cells and batteries, and nano-plastic packaging to enhance food quality, could be in the pipeline according to Sirirurg Songsivilai, executive director of the state-run National Nanotechnology Center (Nanotec).

Nanotec had set up seven associate centers in universities in Thailand with about 400 researchers in total. To attract new researchers—the strategic plan aims for 100 a year up until 2013—the center has so far awarded about 100 scholarships to students to study up to the doctorate level overseas, and 200 more in the country. Dr. Dutta plans on expanding nano education to primary and secondary schools for the next phase of the Initiative.

Preparation for Nanotechnology in Developing Nations

In 2005, the International Center for Science and High Technology (ICS)—United Nations Industrial Development Organization (UNIDO) held a *North-South Dialogue on Nanotechnology: Challenges and Opportunities*[3] in Trieste, Italy.

The workshops were challenging, and education was one of the key elements for consideration in developing nations.

The following recommendations for development of human resources were considered:

- Find the needs of the current industries in each developing country to establish plans for future nano products. Search for the supply and demand side of prospective niche products.
- Translate English teaching materials into regional languages and/or build up the students' English proficiency to follow courses in nanotechnology.
- Develop informational workshops to introduce and address virtues of nanotechnology to industry.

Conclusions regarding universities:

- Retrain faculties in universities to enable them to teach nanotechnology through short-term and specialized courses.
- Identify overlaps amongst faculties, learn more about integrated subjects, and develop inter-and intradepartmental collaborations in universities.
- Utilize a lecture/research combination for nanotechnology training.

Conclusions regarding collaborations:

- Create a network for the exchange of students among universities in Asia-Pacific, as well as more developed nations in Asia, Europe, and America, so that in the final year of master's courses students can make use of the vast available infrastructure of the advanced countries to achieve hands-on learning experiences.
- Collaboration on textbook development, preferably made available over the Internet, for easier dissemination of information regardless of the buying power of students.
- Expanded modules of current syllabus to easily add topics specific to nanoscience information in biology, chemistry, physics, mathematics, and engineering.

The benefits of retraining engineers as an example to consider:

- Retraining engineers from all disciplines who were already in the workplace before the first Nanotechnology Initiatives were funded will be most effective. Job losses in the semiconductor industry have led to a large, available engineering workforce that is well distributed among the major high-technology regions. Since most semiconductor engineers are well acquainted with the basics of solid-state physics, it is very easy for them to make the transition from semiconductor engineering to nanotechnology. Computer chips and storage media are already down to the 32-nanometer size. Mobile devices are demanding smaller chips with more memory and sensors with more power each year. Engineers in the automotive and aerospace industries also need retraining in composite materials for development and production. For corporations with in-house education programs, retraining engineers will be faster and more productive.

New Courses for Aerospace and Aeronautics Engineering Professionals

CANEUS[4] Micro-NanoTechnologies for Space Applications (MNT) Initiates Education Programs

CANEUS Europe is initiating MNT Training Centers and a Summer School Program for Aerospace Professionals. The hands-on Training Centers and Summer School will provide professionals with experience in developing the latest in MNT products and systems. As we discussed in the workforce training chapter, the engineers at Boeing trained in metallurgy were finding it necessary to learn how to work with nanocomposite materials, which have different properties than metal. Partnering with governments, universities, and industry, CANEUS International is moving forward with training for professionals in aerospace and aeronautics.

Creating a Pipeline for Emerging Technologies

Milind Pimprikar, chairman for the CANEUS (Canada-Europe-US-Asia) organization he founded in 2002, presents a first-of-its-kind approach to address the problem of transitioning new and emerging technologies through the infamous "Valley of Death" stage. Noting the lack of a robust mechanism to take new concepts from proof-of-principle stage to system-level implementation, CANEUS has spearheaded the creation of a smoothly functioning technology development "pipeline." Namely, bringing all of the stakeholders together: the inventors, system developers, end-users, and investors. CANEUS' novel approach creates a three-year system for prototype development. These projects will transition laboratory-scale ideas into commercialization and production. In the case of space technologies, it is important that these prototypes have spaceflight testing in extreme environments to qualify for infusion in space missions. The program enables new technologies (materials, devices, and subsystems) to gain the necessary spaceflight experience during early stages of development, rapidly, and at low cost, via the creation of international collaborative sector consortia. These international public/private partnerships between industry, university, and government stakeholders pool membership's resources to define and execute high-risk, high-cost projects and initiatives. Sector consortia thus constitute smoothly functioning development "pipelines" for emerging micro-nano-technology concepts such as: advanced materials, next-generation devices and systems, reliability and testing technologies, fly-by-wireless technologies, and small satellite systems.

Over the past eight years, CANEUS organized biennial conferences and topical workshops globally in partnership with organizations such as IEEE/ASME/SPIE and NASA to bring this niche area of training to the forefront for collaboration. The expansion of space agencies during this time period in Europe, Asia, and South America prompted the group to develop professional education programs that address emerging micro/nanotechnology for space programs.

Working together, CANEUS Europe and CANEUS Asia launched the "Training Centers and Summer Schools" program in Italy, Spain, and other parts of Europe as follows:

Rome, Italy

Dr. Coumar Oudea, president of CANEUS Europe and Milind Pimprikar, chairman of CANEUS International announced the creation of the *CANEUS Europe Micro-Nanotechnology Centers for Excellence and Summer Schools* focusing on aeronautics, space, and defense applications, that will integrate MNT research, education, and aerospace applications through partnerships with universities, governments, and industries. Dr. Francesco Svelto, of the Italian Space Agency (ASI) hosted the CANEUS Europe Strategic Meeting at the ASI headquarters in Rome to plan focused CANEUS activities in Europe during the 2010–2020 period.

Key CANEUS colleagues representing the nine European countries (Italy, France, Spain, Greece, Germany, Portugal, Switzerland, The Netherlands, Belgium, and Germany) met to formulate CANEUS workshops/hands-on training and Summer School programs in their respective countries. The meeting was convened by CANEUS chairman Milind Pimprikar as part of the CANEUS Europe and CANEUS Asia strategic plan for the 2010–2020 periods.

Spain

Barcelona, Spain: Novel MNT Devices, System on Chip (SOC), and System in Package (SIP)

Headed by Dr. Carles Ferrer, the hands-on MNT Center for Excellence for Aerospace Professionals and program managers will focus on creating novel MNT devices, systems on chip (SOC), and systems in package (SIP) that could, for example, cover one week of theory and the second week for hands-on training for aerospace and defense applications. Additionally, it will also build on the European MNT biomedical activities, enable high-capacity energy and information storage devices, and produce sensors and components for aircraft and marine, as well as other important applications. The first training sessions are planned for April 2011 and September 2011 and will be limited to 20 industrial participants. Additionally, Barcelona will also host the MNT for Aerospace and Defense Summer School in 2011/2012.

Italy

Capua (Naples), Italy: Nanomaterials for Thermal Protection

Headed by ASI (Italian Space Agency), the CANEUS hands-on Training Center for Excellence will focus on nanomaterial concepts for hypersonic space and defense applications. Again, the two weeks theory and hands-on program will be limited to 20 participants with the first session planned for 2011/2012.

Frascati (Rome), Italy: Nanosensors and Materials for ICT, Aerospace, Biomedicine

Dr. Stefano Bellucci, of the National Institute of Nuclear Physics will head the CANEUS hands-on training center on nanotechnology (nanomaterials and sensors) and Summer School for aerospace and biomedical applications. The training and Summer School program, limited to 20 professionals, will build on the EU project CATHERINE on nano-interconnections as well as "Innovative Methodologies" for the risk assessment from occupational exposure to nanomaterials.

Greece

Greece: CANEUS Workshop on FBW (Fly-By-Wireless)

Headed by Dr. Constantin Papadas, Greece will host a CANEUS Workshop on FBW for Aerospace applications planned towards early 2011.

Belgium

UCL–Belgium: CANEUS Workshop on HE (Harsh Environment) Sensors

Headed by Microsystems Chair Prof. Laurent Francis, UCL–Belgium, will host a CANEUS Workshop on MNT Harsh Environment Sensors for Aerospace and Defense Applications planned towards 2010 or early 2011. The two- or three-day workshop is limited to 100 participants and will help advance the concepts presented at the CANEUS-ESA International MNT for Space Week, and also formulate funding proposals for the EU framework program.

India

International Academy of Astronautics (IAA) and CANEUS International Plan Joint Workshop in India

The joint CANEUS-IAA workshop is expected to be the first in a series of projects/initiatives the two organizations will pursue together in the coming years. The joint programs are focusing on the CANEUS consortia model based on open innovation to provide the space industry with access to the emerging micro-nano-technologies worldwide.

European Space Agency (ESA) Highlights Online Games As Key Future Technology

Video gaming has become one of the globe's most popular pastimes. Fans say games are often educational; their detractors answer they are anything but. Might ESA have something to learn from gaming? A new Agency study says the answer is yes.

It comes from ESA's Technology Observatory, which is tasked with scanning nonspace sectors to look for developments with potential for spin-in or joint research. The study, Online Game Technology for Space Education and System Analysis, looks at potential applications of different online game-playing technologies from the simplest content-oriented games to massively multiplayer online (MMO) virtual worlds.

The study highlights a number of ways in which these technologies could benefit ESA aims: immersive environments based on these technologies could enhance collaborative working of project scientists and engineers. It was also recognized that exciting online games could prove an excellent tool for promoting space and supporting the teaching of science, technology, engineering, and math.

As part of the study, a video of a potential future game environment was produced showing future human exploration of Jupiter's ice moon Europa.

Exploratory Learning Environments

Secondary school and university students are considered as the natural target audience of such "exploratory learning environments," being already familiar with the interaction principles involved. But other important groups are also recognized: educators, members of the public without any previous interest in space, space professionals, parents of students and, of course, current games players.

The study shows that games could be valuable to educators who want to learn more about the space industry and be in a better position to support students. Simulation games designed as information tools and meeting places could be used to inform and attract potential future space professionals.

Learning through Games

ESA would need to establish partnerships with other stakeholders before the use of online games technology becomes practical reality. The study recommends looking into public–private partnerships, which would be self-financing, enabling online games operators to invest in development, support, and game content, building upon technical and scientific expertise provided by ESA. Widespread consultation concerning the design and promotion of any potential product would be required for such an initiative to become a successful educational tool. ESA experts and representatives would need to involve parents and educators, national space agencies, and industrial contractors.

Study Background

The study has been undertaken by MindArk PE AB of Sweden, a company providing technology platforms for the development and operation of MMO games. Funding was provided through ESA's General Studies Program (GSP) and the study initiated by the Software Systems division of ESA's Technical and Quality Management Directorate.

The GSP interfaces in various ways with all ESA programs, but its main role is to act as a "think-tank" to lay the groundwork for future Agency activities. By proposing this study within GSP, Software Systems sought to investigate the potential use of advanced software technology and platforms in order to prepare for the next generation of engineers entering the space business.

A video preview[5] from Mindark is on their Web site.

NASA MMO[6] Game "Moonbase Alpha"

NASA has been exploring games as education for the past decade and is hoping to create a very popular online gaming/educational experience that will not only entertain, but interest young people in careers in science and engineering.

The NASA Learning Technologies (LT) at Goddard Space Center has initiated the project after studying gaming environments since 2004. They have found that synthetic environments can serve as powerful "hands-on" tools for teaching a range of complex subjects. Virtual worlds with scientifically accurate simulations could permit learners to tinker with chemical reactions in living cells, practice operating and repairing expensive equipment, and experience microgravity, making it easier to grasp complex concepts and transfer this understanding quickly to practical problems. MMOs help players develop and exercise a skill set closely matching the thinking, planning, learning, and technical skills increasingly in demand by employers. These skills include strategic thinking, interpretative analysis, problem solving, plan formulation and execution, team-building and cooperation, and adaptation to rapid change.

The power of games as educational tools is rapidly gaining recognition. NASA is in a position to develop an online game that functions as a persistent, synthetic environment supporting education as a laboratory, a massive visualization tool and collaborative workspace while simultaneously drawing users into a challenging, game-play immersion.

There are concerns that a NASA space reality platform may not be very popular, as other in-space universes offer science fiction or space fantasy, with epic spaceship battles and alien encounters. Hopefully NASA and the MMO developers will strike a healthy balance between education and entertainment. Developing the game with an underlying story to keep the players interested promotes learning while having fun. Student input during development and testing phases would also be wise—because the game has to be fun and challenging from their perspective—not ours.

Nanotechnology Initiatives and Educational Resources around the World

Australia—Access Nano[7]

AccessNano is a unique, cutting-edge nanotechnology educational resource designed to introduce accessible and innovative science and technology into Australian secondary school classrooms.

Teaching Modules[8]

AccessNano provides teachers with 13 ready-to-use, versatile, Web-based teaching modules, featuring PowerPoint presentations, experiments, activities, animations, and links to interactive Web sites. Topics covered fit into current Australian curricula requirements and include teaching units for ages 7 to 11.

Netherlands—Fractal Geometry Program[9]

The Fractal Imaginator is an innovative program for generating fractal images, which are very helpful for measuring wavy surfaces at the nano scale of science. Based on Fractal Zplot, parts of this program have been in continuous development since 1989. New additions to Imaginator are drawn from Jules Ruis' program "Fractal Awareness" algorithms. Fractal Imaginator currently supports Julius Ruis' sets of Julia basins, orbit-traps, level sets, Newton's method, and Phoenix curves, as well as the standard Mandelbrot and "warped" Mandelbrot sets. 3D fractal types include quaternion, hypercomplex, cubic Mandelbrots, cquat, and octonions.

NanoEducation in Russia

Despite the economic recession, the Russian Federation continues investing in nanoeducation. The core point of the program is installation of special tools in educational establishments of different kinds and organization of special training nanolaboratories.

The Federal Directed Program "Development of Nanotech Infrastructure in The Russian Federation," initiated by President Vladimir Putin, was adopted in 2008 and will be active through 2010. Nanoeducation is one of the important parts of this program, which is governed by the Ministry of Education and Science of Russia.

According to the Program, the main activity in nanoeducation for today is to equip the central universities and other educational establishments with training nanolaboratories. These are experimental classes fitted with special microscopes and connected computers, where students can train under the auspices of a professor or lead their own scientific investigations.

In 2008, 35 higher schools in Russia were equipped with special nanoeducational complexes. The universities were the first to try the experiment and were chosen according to their specialization and rating from all over the country.

Professors and students have given favorable reports about the program. "For the first time the students, even on the earlier stages of education, received an opportunity to look at the nanoworld by their own eyes and to imagine the real picture. I am sure that this positive fact will make the educational process much more interesting and clear,"—said the professor of one of the main Russian universities.

At the moment the nanoeducational complexes are being installed in 48 Russian secondary schools. Primarily, these are the schools specializing in physics, situated in different parts of the Russian Federation. At the end of the last year the equipment was delivered and most of the classes are already teaching.

It is worth mentioning that to date the Russian Government is equipping classrooms and has contracted with NT-MDT Co.,[10] a global company producing complex nanotech devices. For Russian schools and universities the company provides special training tools called "NANOEDUCATOR." It is a scientific educational complex with a set of learning aids, accessories for introducing students to nanotechnology and giving them a basic understanding of how to work with objects at nanoscale. It is a student-oriented SPM developed for work by even first-time microscope users.

Turkey—University Makes Leap with Nanotechnology

Sabanci University's Nanotechnology Research and Application Center, or SUNUM, with research priority in agriculture, water and environmental cleanliness, and medicine, is also of interest. Their goal is to become among the world's top ten nanotechnology centers within three years.

Egypt—Nanotechnology Comes to AUC

The American University in Cairo (AUC), through the Yousef Jameel Science and Technology Research Center (YJSTRC), is collaborating in physical sciences, engineering, nanotechnology, and bio-nanotechnology. Research includes novel diagnostic tests for sensitive detection of the hepatitis C virus, detection of cancer biomarkers, nanodevices, smart-brick sensors to analyze building safety and warn of fires and earthquakes, and to mount in and around cars to assist airbag deployment and sense low tire pressure and objects around the vehicle. AUC professors are constructing micro electromechanical (MEMS) and nano electromechanical systems (NEMS) for imaging, communication systems, blood pressure regulation, muscle stimulators, high-density storage media, and lab-on-a-chip.

Egypt—First Nanotechnology Center to Boost Research[11]

Egypt recently launched its first nanotechnology center aimed at boosting the country's technological education and scientific research applications. The project is a collaboration between the Information Technology Industry Development Agency, a government institution, the state-run Science and Technology Fund, and the IBM Corp. Academics from Cairo University, Egypt's largest state-run university, will contribute to the center. Research will focus on the production of solar and renewable energy, water desalination, and modeling of software programs. Planning for the center was based on a memorandum of understanding signed in September 2008 between the development agency, on behalf of the Egyptian Ministry of Communication and Information Technology, the Science and Technology Development Fund representing the Ministry of Higher Education and Scientific Research, and IBM. It is funded by contributions from information technology and communications companies operating in Egypt.

Hungary—Social Network for International Nanoscience Community[12]

The Internet Nanoscience Community, TINC, was created by Hungarian chemistry student Andras Paszternak. It now provides a rich menu of communication tools for the international community of scientists working in the growing field of nanoscience and nanotechnology and recently passed the 3,000 members mark. The virtual nano community is fully equipped with all the functions one expects from a modern online networking site: personal chat, a scientific forum, more than 50 thematic groups, including microscopy, nanomedicine, and even a discussion forum on safety and toxicity. TINC is also a media partner for more than 30 nano conferences on different topics in 2009 and 2010. The Internet Nanoscience Community has pulled together a community with more than 3,100 members, researchers, students, industrial partners from Europe, India, the United States, and 50 other countries. TINC is open to everyone from postdoctoral researchers and professors to students everywhere. "There is only one important assumption: you have to be interested in nano!" adds Paszternak.

Iran—Regional Leader for Nanotechnology[13]

The Islamic Republic of Iran holds first place for nanotechnology in the region and will enter the ten leading world countries by 2015, according to Iranian official representatives. "When the nanotechnology headquarters (in Iran) were established in 2004, we held 57th place in the world. Our goal is to enter the fifteen leading countries in this scientific sphere in 2015," head of the special organization for the development of nanotechnology, the Iranian Nanotechnology Initiative, Said Sarkar, said. There are fifteen Iranian universities which admit students applying for master's degrees in nanotechnology and another five universities admitting students for doctorates. The major goal from headquarters is for Iran to profit and flourish financially with the help of nanotechnology. According to Sarkar, the volume of this market segment will hit a thousand billion dollars in the future. According to plans, Iran will receive 2% of this total. "At present, there are 20 companies already operating and 50 new companies which are developing nanotechnology in Iran," he said.

Iran Opens Nanotechnology Research Center on Agriculture[14]

The Iranian Agricultural Biotechnology Institute, affiliated with the Ministry of Agricultural Jihad, established a nanotechnology research center in a bid to conduct wider studies on the applications of nanotechnology in agriculture. According to the Iranian Nanotechnology Initiative Council, the center is allocated to studies on new methods of nanotechnology-based products' packaging, the invention of pest control methods, and the effects of nanomaterials on the agricultural products and the environment. The cost of equipment needed for the laboratory is estimated at about 1.75 million Euros, which has been totally funded from the credits of the institute, said Dr. Khiam Nekouyee, the chairman of the institute.

Czech Republic Launches New Nanotechnology Research Initiative with EU Support[15]

Toxicology, mechanical engineering, nanoscience, and veterinary medicine are the subjects of four major new EU-funded research initiatives launched in the Czech Republic. The projects, which will share over CZK 2 billion (EUR 77 million), are financed under the Operational Program Research and Development for Innovation (OP R+DI), which receives EU support through the European Regional Development Fund (ERDF).

Serbia Must Invest in Scientific, Technological Development[16]

Deputy Prime Minister and Minister of Science and Technological Development Bozidar Djelic stated that a draft strategy for scientific and technological development is one of the Ministry's key achievements in the first year of its term. Djelic explained that the draft strategy defines their priorities, including biomedicine, new materials, and nanoscience, as well as environmental protection and plans to develop a Center of Excellence for this research. They also provided grants to 250 researchers, scientific training for 1,200 Serbian scientists abroad, RSD 8 million set aside for Ph.D. studies, and RSD 12 million for the Petnica research station. "Serbia must invest in its scientific and technological development as it cannot become a modern and developed country if it exports its best workers," said Djelic, adding that the Ministry has approved the realization of around 500 basic research projects.

Vietnam to Use Methanol Fuel Cells[17]

Based on his successful research on manufacturing direct methanol fuel cells (DMFC), using nanomaterials and nanotechnology being conducted in 2004, Dr. Nguyen Manh Tuan from the Vietnamese Academy of Science and Technology's Ho Chi Minh City Institute of Physics has unveiled different types of fuel cells that use methanol. A fuel cell can light one 20-mW LED bulb for four hours with only 3 mL of methanol, according to the results of the experiment. Dr. Tuan added that this is equal to other cells that are currently being used in developed nations: "The fuel cells have been recognized worldwide as having a long life-span, are easily recycled and using nanotechnology and materials, they cause no harm to the environment." He also felt that Vietnam could supply up to 80% of the materials needed to produce the cell. At present, Tuan and his colleagues are finalizing their study to commercialize this product for use in mobile phones, laptops, buses, and taxis. Meanwhile, globally famous electronic companies such as NEC, Toshiba, LG, IBM, Motorola, Ford, and General Motors have successfully used fuel cells in their products. The world will spend around three billion USD on fuel cells in 2011 as they will help to counter the exhaustion of fuels and help combat global environmental pollution, said Dr. Tuan. The First International Workshop on Nanotechnology and Applications (1st IWNA) was held by the Vietnam National University of Ho Chi Minh City (VNU-HCM) and partners in VUNG TAU City, Vietnam in 2007.

Sri Lanka—Pyxle Develops Nano-Based Information Portal for Sri Lanka[18]

The Sri Lanka Institute of Nanotechnology (SLINTEC) has launched its first Sri Lankan nano-based information portal. This interactive Web portal is designed to act as an information hub for nanotechnology research in Sri Lanka, with a focus on creating awareness of nanotechnology among students, educating potential investors and clients, enabling the government to measure the performance of funding, providing a forum for scientists to share research, satisfying public curiosity, and aiding the business sector decision makers in their planning and evaluation of nanotechnology. SLINTEC science leader, Professor Ajith De Alwis, said, "Our vision is to be the leading sustainable nanotechnology research entity in Asia, and SLINTEC is working towards transforming Sri Lanka into a strong Nanotechnology-focused nation. Our website is a key interface in reaching this vision."

Canada—Nanomaterial Is Biggest Hope for Struggling Forest Industry[19]

The biggest hope for Canada's struggling forest industry could be something very small with a big name—nanocrystalline cellulose. It is nanomaterial—in simplest terms, a product derived through engineering at the molecular scale—derived from wood fiber. Nanocrystalline cellulose, or NCC for short, has yet to make an impact on the marketplace, but in a few years companies could find commercial uses in goods as diverse as lipstick to SUVs because of properties such as strength and toughness, biodegradability, and ability to "tune" colors without dye. And governments and industry are hoping it could be part of a new Canadian bio-economy producing nanomaterials and other products derived from the forest.

Canada has developed Nanoscience Centers in Alberta, Ontario, and Waterloo Universities.

Mexico—Center for Biomedical and Nanomedical Research to be Built by Mexico City Government[20]

The government of Mexico City announced plans to build a world-class center for biomedical and nanomedical research, called Campus Biometropolis. The center will be integrated with the National Autonomous University of Mexico, the top Spanish-speaking university in the world.

Africa

Zimbabwe—President Caps 393 CUT Graduates[21]

The Chinhoyi University of Technology (CUT), Zimbabwe, graduated 393 students in 2009, the fifth program partnered with the University of Cape Town's Center for Materials Engineering in South Africa, that focuses on the development of platinum alloy catalysts using nanotechnology.

Cape Town, South Africa—Rice Connexions[22]*–South Africa Partnership in Open Education*

Houston-based Rice University and Cape Town, South Africa-based Shuttleworth Foundation are jointly developing one of the world's largest, most comprehensive sets of free online teaching materials for primary and secondary school children. Using their open-education projects—Rice's Connexions and the Shuttleworth's Siyavula—the organizations work to transform South African primary and secondary education based on open-access educational content, open-source software, and online educator communities.

Ireland

Nanotech discovery could bring about the end of animal testing,[23] through a sensing system that monitors the effects toxicants have on human and animal cells. The funding for the project was from the FP6 European program and included several European academic and industry partners. They have 600 researchers working in nanotechnology and 300 students undertaking Ph.D. programs related to nanoscience.

Switzerland

Swiss Nano-Cube[24] is the national information, learning, and teaching platform dedicated to the topics of micro- and nanotechnologies (M&NT) for secondary schools, companies, and industry associations. The platform will offer education modules in German and information materials from the area of M&NT to teachers, students, as well as vocational experts from industry. The platform is jointly developed by the Innovation Society (St. Gallen) and the Swiss educational Federal Institute for Vocational Education and Training (Zollikofen).

Malaysia—Government to Establish National Innovation Center

The government of Malaysia will establish a National Innovation Center and a network of Centers of Innovation Excellence as a step toward accelerating national innovation and commercialization activities. It was also decided that nanotechnology development would be given priority and be made one of the resources of the country's new economic model, because it represents a new and advanced technological field.

Philippines—DoST Unveils Nanotech Roadmap[25]

The Department of Science and Technology Philippine Council for Advanced Science and Technology Research and Development (DoST-PCASTRD) announced their 10-year nanotechnology roadmap. They plan to focus on five key industries: IT and the semiconductor, energy, agriculture, medicine, environment and food.

Kuwait—A New National Vision Includes Nanotechnology

> With time pressing us with regard to the use of oil which is finite, it is time to divert our attention to the energies of the future that are renewable and infinite; solar, wind, and wave technologies to name a few. We should also, perhaps, endeavor to transform our barren desert into an oasis not for any other reason than to prove that it can be done. We should work towards ensuring the development of new national industries in fields as diverse as medical technology, nanotechnology, underwater exploration, flight, agriculture, as well as into the chemical industry to name a few possible tracks.[26]

Brazil—Brazil's Petrobras Throws a Half-Billion Dollars at World's Technological Race[27]

Since the passing of the Petroleum Act, in 1997, which added a clause to contracts between concession-holders and the National Petroleum, Natural Gas and Biofuel Agency (ANP) providing for mandatory investment in research and development (R&D), partnerships between Petrobras and the Brazilian technology cluster have increased significantly. Set in 2005, the clause stipulates that at least 1% of gross revenues from oil fields in which Special Participation is due must be invested in R&D. Out of that figure, 50% must go to national science and technology institutions. Since then, 38 thematic networks have been created, and the most competent national institutions in their own segments have been invited to them. According to Petrobras, the networks include themes such as an increase in production of heavy oil, research for new materials for the refining process, and nanotechnology applied to the energy industry for the development of bio-products.

Cuba—Cuban Scientists Obtain Their First Images of Atoms

Researchers at the Science and Technology of Materials Center of the University of Havana recently obtained images of atomic resolution with the first nanotechnology instrument made in Cuba.

The developers of the project told *Juventud Tecnica* magazine's digital edition that the scanning tunneling microscope (STM) was made by Cuban researchers with the collaboration of the National Autonomous Universidad of Mexico (UNAM) as the first step toward nanotechnology development in the country.

Global Resources for Nanoscience Education

Australia—In2science[28]

Starting in 2004 as a joint venture between the Faculties of Science at La Trobe University and The University of Melbourne, In2science was joined by Monash University in 2008.

In2science currently has partner schools in both the metropolitan and regional areas of Victoria. Within these schools there are university science and mathematics students who volunteer their time to work in the classroom for a few hours each week. Here they are role models for the students, inspiring them to raise their aspirations and achieve their potential in these subjects.

In2science has four primary aims:

- To generate enthusiasm for science (especially the enabling subjects of chemistry, physics, and mathematics) in students in the middle years of their secondary education (grades 7–10)
- To place university students in schools to act as positive role models to secondary school science students, inspiring them to achieve their potential
- Through the role models, promote the value and rewards of science as a positive career choice
- To foster links between schools and universities

In2science draws on the successful *STAR*[29] program at Murdoch University, WA.

Germany

NanoReisen[30]—Nano Journey, Adventures beyond the Decimal—Middle School to Adult

NanoReisen takes you on an interactive video trip from the world of matter to the nano-cosmos.

NanoTruck[31]—High-Tech from the Nanocosmos

Sixty exhibits, more than half of which are interactive, a laser show, film showings, multimedia presentations, and an area reserved for experiments: the mobile exhibition on board the nanoTruck road-show vehicle is the central element of the new "nanoTruck—High-Tech from the Nano-Cosmos" campaign designed by the Federal Ministry of Education and Research (BMBF). This exhibition has been designed to bring the topic of nanotechnology directly to the people in an informative and entertaining manner. The main exhibits are so-called "hands-on science" objects, which are designed to help the general public to understand the way in which nanotechnology functions.

United Kingdom

U.K. NanoMission™[32]—Learning Nanotechnology through Games

A cutting-edge gaming experience which educates players about basic concepts in nanoscience through real-world practical applications from microelectronics to drug delivery.

The aim is to inspire youngsters around the world about nanotechnology, opening their eyes to choosing it as a career. There are presently four modules available free to schools around the world.

Developed by PlayGen,[33] leader in serious games for education.

U.K. Nanotechnology for Schools[34]

When you think of building, you always think of pylons, bricks, cement, glass, scaffolding, cranes, bulldozers, or heavy industrial machinery. But this is not always the case. New methods of building are being developed that do not use girders or cement, but individual atoms. Engineers can now manipulate single atoms to create tiny machines and things that can be used in new and exciting inventions.

Can you imagine a robot the size of a blood cell that swims around in your bloodstream, detecting and destroying cancer cells? Can you imagine a supercomputer that fits into an earring? Both of these wild ideas and more may become reality in the next 15 to 30 years! Nanotechnology is the study of tiny particles that would be used for these constructions, and how to make the constructions themselves, and what properties and potentials these machines could have.

U.K.—The Vega Science Trust Videos[35]

This Web site features videos of interviews with scientists. The Nobel Prize Laureate Richard P. Feynman, Physics, was filmed in a set of four videos titled *The Next Big Thing (Nanotechnology)*, you can stream in your classroom. Teachers and students can explore the library of lectures and resource videos for the classrooms.

U.K.—Espresso Education[36]

The leading digital curriculum service in the United Kingdom. Featuring inspirational curriculum resources, Espresso provides an extensive library of high-quality, video-rich, broadband teaching resources and student activities that motivates pupils and supports teachers.

Understanding Nanotechnology[37]

This is an informal information Internet Web site dedicated to making nanotechnology concepts and applications understandable by anyone. On this site are excerpts from *Nanotechnology for Dummies* (Richard Booker and Earl Boysen, Wiley Publishing); explanations of nanotechnology concepts; links to articles and resources, such as manufacturers' Web sites; and nanotechnology stories in the news.

Nobel Prize Educational Games[38]—Grades K–12

You do not have to be a genius to understand the work of the Nobel Laureates. These games and simulations, based on Nobel Prize-awarded achievements, will teach and inspire you while you are having FUN!

Switzerland

Switzerland—École Polytechnique Fédérale de Lausanne (EPFL)

The BioWall constitutes a major step toward the creation of intelligent, bio-inspired electronic tissues, capable of evolving, self-repairing, self-replicating, and learning. In its current form, the BioWall surface combines the possibilities offered by the very latest information technology with the most instinctive of human gestures—touch.

The current BioWall:[39] You can see the BioWall operating live through EPFL's Webcam, Monday to Friday, 8:00–19:00 GMT +1

Switzerland—Original Virtual Nano Lab[40]

Visit the Original Virtual Nano Lab located at the University of Basel, Switzerland. First demonstration (April 2002) of the virtual nanoscience laboratory: The specific subject contents of the learning basis are elaborated in close cooperation with the newly established National Center of Competence and Research in nanoscale sciences in Basel. The development of virtual basis establishes a new era of didactics and methods. Europe is exploring the e-learning age; Swiss Virtual Campus is a pioneering program and makes an important contribution to the education in the twenty-first century.

Egypt—In2nano High School Club[41]

On the road of life you lead, or follow, or get out of the way. In2nano, the NanoClub for Egypt's aspiring teens is a well-planned attempt to put them in the lead. The in2nano project is designed to prepare high school students for the sweeping changes that nanoscience and nanotechnologies are bringing to the world.

Why nano? Nanoscience is an emerging field that has made revolutionary scientific breakthroughs in recent years. The technologies resulting from the nano-revolution will have far-reaching societal implication across all aspects of life including healthcare, agriculture, energy, water, and the environment. To keep up with the rapid pace of these newly developing technologies, we need a high degree of scientific literacy in the Arab world.

The in2nano project aims to ignite the minds of Egyptian youth to imagine how technology and science can offer solutions for their future. The project consisted of both an "actual" and a "virtual" classroom. The first half consisted of a series of lectures to be delivered to participating public and private schools across Egypt in 2009. The second half of the equation is the Web site, which relays information about the program itself and is an information portal for students, their caregivers, and educators. Finally, a FunDay was organized to bring the learning of nanoscience to life.

In2nano aims to help Egyptian students develop a passion for science, but also for knowledge and creativity more generally. By emphasizing teamwork and creativity students will learn to reach beyond the limitations of yesterday into a brighter tomorrow.

Nanoscience is a new science for a new generation. The future is theirs, we have to help them reach it.

Bulgaria—Nanopolis—A World of Knowledge[42]

Nanopolis is a provider of multimedia education in nano-biotechnology and nanotechnology, mirroring in real time the scientific and technological progress in the field of matter exploration at the atomic scale. Transposing the interaction between research, industry, and education into vivid, interactive, pedagogical multimedia since 1998, Nanopolis currently offers more than 3,000 screens through its online and offline resources.

Developers/Teachers

Multimedia animations explaining nanotechnology concepts are featured at Nanopolis, the e-collaborative multimedia producer that is exclusively dedicated to expanding the nano-biotechnology knowledge base. It offers more than 10,000 scientific validated animations and e-learning pages, available online (at their Online Multimedia Library) and offline (though the multimedia encyclopedic series, in CD-ROM or download format).

FIRST Robotics and NASA[43]—The Robotics Alliance Project Now in 33 Countries

Set up under the sports and entertainment model to develop Superstars in Science, Technology, and Engineering, FIRST celebrates innovators and thinkers, challenging them with robotics as an appealing and fun sport, encouraging the next generation of Superstars that may choose a career in science, technology, and engineering. Each year teams across the world, ages 9 to 14, are assigned a project to complete in addition to building and programming a robot. The theme for 2007 was on nanotechnology and was very successful.

First Robotics Competition Regional Webcasts 2007[44]

Resources for the FIRST kids introduced to nanotechnology in the Lego group were provided by:

- University of Wisconsin Interdisciplinary Education Group
- Exploring the Nanoworld with LEGO® Bricks[45]

The purpose of this Web site and booklet is to show how various physical and chemical principles related to nanoscale science and technology can be demonstrated with LEGO® models. Three-dimensional models are excellent tools for grasping structure-function relationships.

There are a number of reasons to consider using LEGO® bricks for this purpose. First, many people are familiar with LEGO® bricks, and most models can be built with a level of mechanical sophistication that does not intimidate or frustrate the user. Second, LEGO® bricks typically have many connection points, allowing tremendous flexibility in the structures that can be built. A set of bricks can be used to model structures of matter and the techniques used to study them.

Funding is provided by National Science Foundation, through the University of Wisconsin–Madison Materials Research Science and Engineering Center (MRSEC) for Nanostructured Materials and Interfaces, and by LEGO® Dacta®.

Carnegie Mellon University—The National Robotics Engineering Center

Robotics Academy Outreach Programs[46]

With the motto "We're building engineers one child at a time," Carnegie-Mellon provides K–12 Robotics Education, Teacher Training, Robocamps, competition, clubs, and classes as educational outreach.

NASA Classroom of the Future Program[47]

The Classroom of the Future™ (COTF) program is helping to bridge the gap between America's classrooms and the expertise of NASA scientists, who have advanced the frontiers of knowledge in virtually every field of science over the last forty years. The COTF program is administered by the Erma Ora Byrd Center for Educational Technologies™ at Wheeling Jesuit University in Wheeling, West Virginia.

NASA—Welcome to the Space Place![48]

NASA invited you to "come on in" and check out games, animations, projects, and fun facts about Earth, space, and technology.

MIT Global Access Laboratories[49]

Remote Microelectronics Laboratory

Prof. Jesus del Alamo and his team are making a real microelectronics testing laboratory available online to students from anywhere, 24 hours a day. This will revolutionize science and engineering education by providing greater access to state-of-the-art labs, including at other institutions.

Frank Potter's Science Gems—Engineering[50]

This site has some science curriculum for high school and will provide new curriculum for grades 6–9 when it becomes available. The Web site features a number of engineering subcategories that may be of interest to students.

MIT's OpenCourseWare[51]

A free and open educational resource for faculty, students, and self-learners around the world. OCW supports MIT's mission to advance knowledge and education, and serve the world in the twenty-first century. It is true to MIT's values of excellence, innovation, and leadership.

Foresight Nanomedicine Art Gallery[52]

Here you will find a small but growing collection of visual artwork describing many different views of how medical nanorobots and other nanomedical devices and systems might appear. Some of these works have been borrowed with permission from already-published print-media or electronic-media works. Other contributions are original graphics created by the named individual artists especially for this gallery exhibition for your additional enjoyment. The images in the Nanomedicine Art Gallery are organized into three nonexclusive conceptual groupings—Nanorobot Species, Medical Challenges, and Individual Artists—for easy browsing.

Nanoscience Instruments[53]

Nano Science Education[54]

Nanoscience Instruments provides portable desktop scanning tunneling microscopes (STMs) and atomic force microscopes (AFMs) as desktop scanners from NanoSurf for ease-of-use, mobility, and affordable pricing that allows you to easily set up teaching labs and outreach programs. The Web site also has nanoscience curriculum, books, software, an animated gallery, and teacher demos.

The Virtual Nanolab at the University of Virginia was designed with these tools for developers.

Teach Nano[55]

TeachNano seminars provide a unique opportunity for those involved in teaching nanoscience to learn about other nanoscience educational programs and to share experiences. The program includes short talks, hands-on demonstrations, and open discussions.

Free Software for Simulation

NanoEngineer-1™ is for Everyone[56]

NanoEngineer-1 is not just for people with powerful supercomputers at large universities. Actually, NanoEngineer-1 was created for anyone with a personal computer, including MAC and Unix that you can download for free. NanoEngineer-1 is an open-source (GPL) 3D multiscale modeling and simulation program for nano-composites with special support for structural DNA nanotechnology. It features an easy-to-use interactive 3D graphical user interface for designing and modeling large, atomically precise composite systems.

Open Source Tools for Structural DNA Nanotechnology (SDN)[57]

Nanorex is developing open-source computational tools to support research in structural DNA nanotechnology (SDN). To do this, they have extended the existing application, NanoEngineer-1, to provide a foundation of tools for visualization, modeling, and manipulation of DNA, and to make it a framework that can support and integrate other computational tools developed by the SDN community. This work is part of the broader mission to support the development of advanced nanosystems. They have entered a process of collaborative development that will help the SDN research community integrate its diverse tools for design, modeling, and analysis, making them more useful and more widely available. Working with the rest of the SDN community, they would like to make this process serve everyone's needs at multiple levels, both as users and as developers.

Affordable Interactive 3D for the Classroom[58]

At a time when employment headlines seem to go from grim to grimmer, there is one area within the workforce that has and will continue to have an expanding demand. That workforce encompasses a variety of job titles involved in the creation of digital content for interactive visualizations, simulation-based learning, serious games, games for learning, etc. Workers are needed—not only to create such interactive digital content for education but also such content as being demanded by industry.

The Games-to-Teach Project[59]

This is a partnership between MIT and Microsoft to develop conceptual prototypes for the next generation of educational media for math, science, and engineering education. Directed by MIT's Program in Comparative Media Studies, Games-to-Teach is funded as a part of Microsoft iCampus and supported by the Learning Sciences and Technologies Lab at Microsoft Research. The project has grown into the Education Arcade. Two white papers on games for education are available.

Math World—Special Programs[60]

Wolfram Research sponsors both the academic and the corporate communities with direct contributions to education-related programs and scientific research. These programs range from the High School Grant Program, which encourages teachers to explore new teaching methods and develop computer-based classroom materials using Mathematica, to the Collaborative Research Opportunities Program, which offers researchers from universities, laboratories, and other organizations the opportunity to contribute their expertise to collaborative research projects. The Student Intern Program recruits talented students who would like to gain real-world experience and offers internships in all departments of the company each summer.

Curriki Curriculum Development Site for Teachers[61]

Curriki is where all of us—our community of educators, parents and students—can work together to develop interesting, creative, and effective educational materials that the global educational community can use for free. Their goal is to develop curriculum through community contributors, deliver the curriculum globally, and determine the impact by project and by individual. The initial focus is on K–12 curricula in the areas of mathematics, science, technology, reading and language arts, and languages.

Check out the Focus on Science Section.[62]

PG Online Training Solutions[63]

eTraining is an affordable and easy-to-use learning management system that uses the Internet to give students remote access to knowledge, immediate assessment of their understanding, certification of achievement, and peer guidance as required. Students can register and learn at their own convenience, using the e-training site to supplement instructor-led classes.

eTraining software technologies have made it possible for educators, administrators, and learning service providers to create, launch, and evaluate learning over the Internet using nothing other than a standard browser. No programming skills are required to fill an eTraining-based site with existing courses.

References*

1. Overview: Human Resources Development in Nanotechnology 2003 http://www.tntg.org/documents/52.html
1. http://www.faculty.ait.ac.th/joy/new/
2. http://www.scidev.net/en/new-technologies/thailand-nanotech-plan-moves-ahead.html
3. http://www.tntg.org/documents/52.html
4. http://www.caneus.org/
5. http://www.esa.int/esaCP/SEMHGBFKZ6G_index_0.html
6. http://ipp.gsfc.nasa.gov/mmo/
7. http://www.accessnano.org/
8. http://www.accessnano.org/teaching-modules
9. http://www.fractal.org/
10. http://www.ntmdt.com.
11. http://www.universityworldnews.com/article.php?story=20090724101625342
12. http://www.nanopaprika.eu/
13. Source: *Turkish Weekly*—a USAK Publication. USAK is the leading Ankara-based Turkish think-tank. http://www.turkishweekly.net
14. http://english.farsnews.com/newstext.php?nn=8805190901
15. http://www.azonano.com/news.asp?newsID=14952
16. http://www.emg.rs/en/news/serbia/94240.html
17. http://english.vietnamnet.vn/
18. http://www.pyxle.net

* All links active as of August 2010.

19. http://www.edmontonjournal.com/Technology/Nanomaterial%20hope%20forest%20industry/1728825/story.html
20. Source: Gobierno del Distrito Federal.
21. http://allafrica.com/stories/200911300032.html
22. http://cnx.org/. Connexions: Create Globally, Educate Locally
23. http://www.siliconrepublic.com/news/article/14552/randd/nanotech-discovery-could-bring-about-the-end-of-animal-testing
24. http://www.swissnanocube.ch/
25. http://newsinfo.inquirer.net/breakingnews/infotech/view/20091029-232939/DoST-unveils-nanotech-roadmap
26. http://www.kuwaittimes.net/read_news.php?newsid=MTQwMjQ4NzU0
28. http://www.latrobe.edu.au/in2science/
29. http://about.murdoch.edu.au/star/intro.html
30. http://www.nanoreisen.com/english/index.html
31. http://www.nanotruck.de/en/
32. http://www.nanomission.org
33. http://www.playgen.com
34. http://www.nanoscience.cam.ac.uk/schools/nano/index.html
35. http://www.vega.org.uk/
36. http://www.espresso.co.uk/index.html
37. http://www.understandingnano.com/
38. http://nobelprize.org/educational_games/
39. http://lslwww.epfl.ch/biowall/
40. http://www.nano-world.org/en
41. http://www.in2nano.sabrycorp.com/conf/in2nano/08/
42. http://www.nanopolis.net/
43. http://www.usfirst.org
44. http://robotics.nasa.gov/events/webcasts/regionals_2007.php
45. http://www.mrsec.wisc.edu/Edetc/LEGO/index.html
46. http://www-education.rec.ri.cmu.edu
47. http://www.cet.edu/?cat=cotf
48. http://spaceplace.nasa.gov/en/kids/
49. http://www.mtl.mit.edu/users/alamo/weblab/index.html
50. http://www.sciencegems.com/engineer.html#10
51. http://ocw.mit.edu/OcwWeb/web/home/home/index.htm
52. http://www.foresight.org/Nanomedicine/Gallery/index.html
53. http://www.nanoscience.com/index.html
54. http://www.nanoscience.com/education/
55. http://www.teachnano.com/teach/
56. http://nanoengineer-1.com/content/index.php?option=com_content&task=view&id=33&Itemid=34
57. http://nanoengineer-1.com/content/index.php
58. http://www.eonftp.com/lp/3d_education.html
59. http://www.educationarcade.org/
60. http://mathworld.wolfram.com/
61. http://www.curriki.org/xwiki/bin/view/Main/
62. http://www.curriki.org/xwiki/bin/view/Main/FocusOn_Science
63. http://www.elmspro.com/etraining/

Section IV

Framework Applied

> For a successful technology, reality must take precedence over public relations, for Nature cannot be fooled.
>
> **Richard Feynman**
> *U.S. educator and physicist (1918–1988)*

The first chapter in this section reviews the role of federal and state governments and new initiatives to include nanoscience resources into our schools as stakeholders make decisions "from the top down." The following two chapters provide supportive examples and resources for teachers and communities to take the next step in rethinking education as stakeholders "from the bottom up."

13

Assessing the Options for Action and Implementation

> Ethical axioms are found and tested not very differently from the axioms of science. Truth is what stands the test of experience.
>
> **Albert Einstein**

Where Do We Start?

Assessment of the options for implementation of instructional materials requires a review of current National and State Standards required for curriculum development to understand our options for inclusion.

The National Nanotechnology Infrastructure Network (NNIN)[1] is an NSF-funded partnership of 13 university user facilities openly available for research and development in nanoscience. They provide tools and training to support the nanotechnology research needs of researchers in academic, industry, and government. The NNIN Education Office is housed at Georgia Institute of Technology and coordinates education efforts across all 13 sites and with other national programs. NNIN has developed, and anticipates continuing development, of instructional materials for K–12 schools and teachers. The curriculum that has been provided was developed following the current National Standards, along with the State Standards where the centers were located.

Why Aren't the Teachers Using the Resources?

As stated in *NSF 0654-5 Appendix2—A Nanoleap—Part VII: Lessons Learned:*[2]

> Teachers are interested in nano but don't know where it fits into the curriculum. They are also interested in all levels of nano instructional materials from short little introductions/"fillers" to full scale lessons tied to the National Standards:

Nanotechnology and the National Science Education Standards[3]

Inclusion of nanotechnology in the science curriculum will foster interdisciplinary explorations of science in the K–12 curriculum. Nanotechnology uses science on the nanoscale, which occurs at the scale of atoms and molecules. Because nanotechnology is an emerging interdisciplinary field, it can be included in physical science, chemistry, physics, biology, environmental sciences, and engineering. Nanotechnology provides connections between and among the sciences that will help students to develop an understanding of the relationships between disciplines. Many teachers have questioned where this field fits into national and local science standards and how something that occurs at the atomic and molecular level can actually be addressed in the K–12 science curriculum. In general, nanotechnology "fits" into current science curriculum because it relies on numerous science concepts and processes which are part of the National Science Education Standards. Summarized below is where nanoscale science can be addressed in the National Science Standards:

Science Content Standards K–4: At this level, students may not be able to understand the concepts of atoms and molecules, but they are developing the foundation for more advanced understanding of science in middle and high school. Important foundations to be addressed at this level include addressing the following standards:

- **Science as Inquiry** including ability to do and understand scientific inquiry
- **Physical Science** including properties of objects and materials, position and motion of objects and light, heat and electricity
- **Life Science** including organisms and environments
- **Science and Technology** including distinguishing between natural and man-made objects, ability of technological design, and understanding about science and technology
- **Science in Personal and Social Perspectives** including changes in environments and science and technology in local challenges
- **History and Nature of Science** including science as a human endeavor

Science Content Standards 5–8: At this level, students begin to understand science concepts at a higher level and are capable of performing experiments/seeking information to understand these concepts. While middle-level students are still not ready to understand the world of atoms and molecules, they do develop a knowledge about the characteristics of materials which is an important component in understanding nanoscale science. Curriculum materials can include information on nanotechnology that will address the following standards:

- **Science as Inquiry** including ability to do and understand scientific inquiry
- **Physical Science** including properties and changes of properties in matter and light as well as transfer of energy
- **Life Science** including structure and function in living systems, regulation and behavior, and ecosystems
- **Science and Technology** including ability of technological design and understanding about science and technology
- **Science in Personal and Social Perspectives** including populations, resources, and environments, risks and benefits, and science and technology in society
- **History and Nature of Science** including science as a human endeavor, nature of science, and history of science

Science Content Standards 9–12: At this level, students develop a rich knowledge base about the physical and life sciences. They expand upon the knowledge learned in earlier grades to understand the microscopic structures of materials and substances. Chemical interactions, the chemical basis of life, and cell structure and function all become part of the curriculum. Nanotechnology can be included in a variety of high school curricula because of its interdisciplinary nature and its relationship with the basic science concepts taught in grades 9–12.

- **Science as Inquiry** including the abilities to do and understand scientific inquiry
- **Physical Science** including structure of atoms, structures and properties of matter, chemical reactions, properties of light, conservation of energy and increase in disorder, and interactions of energy and matter
- **Life Science** including the cell, matter, energy, and organization of living systems, behavior of organisms, and molecular basis of heredity
- **Science and Technology** including ability of technological design and understanding about science and technology
- **Science in Personal and Social Perspectives** including natural resources, environmental quality, natural and human-induced hazards and science and technology in local, national, and global challenges
- **History and Nature of Science** including science as a human endeavor, nature of scientific knowledge, and historical perspectives

A second level of standards was developed by The American Association for the Advancement of Science (AAAS) Project 2061's *Benchmarks for Science Literacy (1993),* published as the *Atlas of Scientific Literacy,* that increases the levels of complexity for curriculum development. The AAAS Project 2061 focuses their effort to improve science literacy in America through research and development of K–12 curriculum assessments and instructional materials. Project 2061 staff use their expertise as teachers, researchers, and scientists to help make literacy in science, mathematics, and technology a reality for all students. They have scheduled a series of workshops[4] for teachers and curriculum developers in 2010 to define the benchmarks, or you can visit their *Benchmarks on Line.*[5]

A second Web site that references the *Benchmarks* is the online *NSDL Science Literacy Maps,*[6] a tool for teachers and students to find resources that relate to specific science and math concepts. The maps illustrate connections between concepts and also show how concepts build upon one another across grade levels. Clicking on a concept within the maps will show National Science Digital Library (NSDL) resources relevant to the concept, as well as information about related AAAS Project 2061 Benchmarks and National Science Education Standards.

Project 2061[7] began in 1998 by providing an evaluation of middle grades science textbooks using its rigorous curriculum-materials analysis procedure. They represent a strong national consensus among educators and scientists on what all K–12 students should know and be able to do in science. Researchers and materials developers analyze how well curriculum, instruction, and assessment support student achievement of specific learning goals. However, in many states teachers are totally responsible for the development of science curriculum and experiments for their students based solely on national and state standards.

A third level consists of State Standards that carry the strongest influence on textbook publishers, with most states adopting the Texas State Standards and Texas Essential Knowledge and Skills (TEKS) testing since 1998. However, that may change as more states move toward a digital e-learning matrix and provide laptops to their students, eliminating standard textbooks. Over the past few years quite a few states have been experimenting with new technology in the classrooms, such as laptops and electronic white boards that connect to Internet resources. State budgetary constraints may enhance this new direction as textbooks and testing have become increasingly expensive.

Revised Science Standards Would Support President Obama's Challenges to Educators

None of the current standards have specified inclusion of nanoscale science, which might be addressed between the National Governors

Association (NGA) and the Council of Chief State School Officers (CCSSO) in the future. They have recently revised the State Standards for reading and mathematics under the new *Race to the Top* program initiated by President Obama.

Stakeholders Gather to Discuss Nanoscience Education

A workshop titled *Partnership for Nanotechnology Education* was held at the University of Southern California, Los Angeles, CA, on April 26–28, 2009, organized by Dr. James S. Murday, USC Chair, and sponsored by the National Science Foundation.[8] The workshop invited stakeholders to explore the educational challenges and opportunities presented by the nanoscale.

The following set of goals were developed for the participants:

- Identify and examine the present status of "nano" education efforts.
- Identify the infrastructure needed to carry out effective "nano" education.
- Lay the groundwork for functional stakeholder partnerships that address the needs and identify the opportunities.
- Identify mechanisms for the partnerships to provide information for the Nanoscale Science, Engineering, and Technology (NSET) subcommittee member agencies, the National Science and Technology Council (NSTC) interagency Education Working Group proposed in the NNI 2009 reauthorization bills, and other interested parties to use in developing funding goals, strategies, and programs.

Priority Recommendations

On the final afternoon of the workshop the participants discussed the recommendations developed in the prior breakout sessions. They identified the following ten items as warranting immediate action.

1. Creation of a NanoEducation Ecosystem:
   - **Finding:** There are numerous groups around the world addressing STEM education, NanoEducation, nanoscale science and engineering research, and nano-enabled technologies. There is an immediate challenge to integrate these various communities. A focal point is needed to identify, validate, and integrate

the many NanoEducation capabilities that presently exist and to assess what is additionally needed.

- **Recommendation:** The NSET, which has representation from about 25 participating Federal agencies, should create a Nanotechnology Education and Workforce working group that will support the agency efforts toward addressing education and workforce issues. An education and workforce-focused consultative board to the NSET should also be created, comprising the various principal stakeholders. National Nanotechnology Coordination Office (NNCO) funds (or other contributions from the various Federal agencies) should be used for this effort.
- **Principal Stakeholders include**: The Executive Office of the President, including the Office of Science and Technology Policy (OSTP) and the National Science and Technology Council (NSTC); the NNI participating Federal agencies; National Education Association (NEA); National Science Teachers Association (NSTA); professional science and engineering organizations (e.g., American Chemical Society (ACS), American Institute of Physics, Society of Manufacturing Engineers, etc.); NanoBusiness Alliance; and STEM Education Coalition.

2. Standards of Learning (SOL):
    - **Finding:** Without incorporating the current understanding of nanoscale science and engineering into science and engineering learning standards in each of the states, action at the K–12 levels of education will be minimal, and increasing nanoscience literacy toward a productive workforce will not be maximized. The National Governors Association (NGA) has approached Achieve Inc.[9] (with whom they developed Benchmarks for English and Math) with the task of preparing common core learning standards in the physical sciences that might be adapted by each state for its own learning standards.
    - **Recommendation 1**: The NSET initiate contact with the NGA, the Council of Chief State School Officers (CCSSO), and Achieve Inc. to foster an effort to appropriately introduce the nanoscale into the common core standards.
    - **Recommendation 2**: Participants in the many U.S. Nano Centers (Part 3: Chapter 9) should begin working with their own State Education Departments toward science learning standard revisions.

There was strong agreement that "nano" needs to be in the K–12 education world; it is not advanced science, but rather simply part of twenty-first century science and is enabling the rapid entry of nano-enabled technologies into the commercial marketplace.

An example of a current standard that could be redefined to include the "nanoscale" in the description where small science is mentioned:

Example from a Current High School Physics Standards of Learning

PH.14: The student will investigate and understand that extremely large and extremely small quantities are not necessarily described by the same laws as those studied in Newtonian physics. Key concepts include:

- wave and particle duality;
- wave properties of matter;
- matter and energy equivalence;
- quantum mechanics and uncertainty;
- relativity;
- nuclear physics;
- solid state physics;
- superconductivity; and
- radioactivity.

In this standard, the teacher is directed to look at "extremely small quantities" but given no guidance as to what that means. Though this standard was written in 2003, nanostructure is not mentioned. A 2003 survey of middle and high school science teachers in Kentucky found that only 33% of those science teachers were familiar with the concept of nanotechnology, and only 60% of the teachers surveyed were even aware of the concept (18% said that they "understood" it). Thus if ALL students are to be introduced to nanoscience as a size/scale, all teachers must know to teach it—something that requires its inclusion in the standards.

- **Principal Stakeholders include**: OSTP/NSTC, NSF, Department of Education (DoEd), CCSSO, NGA, Achieve Inc., NSET's Nanotechnology Public Engagement and Communications Working Group (NPEC), Association of Science and Technology Centers (ASTC), NSTA and the state-based affiliates, International Technology Education Association (ITEA) and the state-based affiliates, American Association for the Advancement of Science (AAAS), and National Research Council (NRC).

3. Teacher Education and Training:

    - **Finding:** There will be growing inclusion of nanoscale science, engineering, and technology into standards of learning. There are also growing learning resources for K–12 audiences that address nanoscale science, engineering, and technology. Teachers will need to be trained to use these resources.

- **Recommendation:** The various NanoCenters can be a vital resource to provide materials, training, and information. They should be encouraged to be more proactive toward K–12 teacher training.
- **Principal Stakeholders include**: NSTA, DoEd, NSF, other Federal agencies whose mission supports teacher training and workforce development, CCSSO, ASTC, and ITEA.

4. K–12 Curricula and Teaching Aids:
   - **Finding:** There are several Web sites with materials that address curricula supplements, teaching aids, and science fair projects. In particular, the NSF-funded Nanoscale Science and Engineering Centers (NSEC) have been very productive at developing innovative approaches to Nano Education. However, the materials are widely dispersed, are of non-uniform format, and have varying degrees of refinement.
   - **Recommendation:** The DoEd, working closely with the NSTA and cyber-oriented curriculum developers, create a central Web site, to house all teacher/student materials categorized by age/grade level. The NSTA should serve as the evaluator for quality control to ensure Web site materials are of high quality, are in a format readily utilized by K–12 teachers, are carefully indexed to the various state learning standards, and can be readily accessed from the NSTA Web site. Additional well-designed, highly interactive, media-rich, online learning tools should continue to be developed.
   - **Finding:** The physical sciences (biology, chemistry, physics) require hands-on experiences as part of the learning process. While inclusion of laboratory work at the local schools is a necessity, some laboratory learning may be beyond the capability and/or budget of local schools and personnel.
   - **Recommendation:** The NSF National Nanotechnology Infrastructure Network (NNIN), NSEC, DoE NanoCenters, and the National Institute of Standards Technology (NIST) Center for Nanoscale Science and Technology work with the NSTA and the DoEd toward the preparation of on-site and/or remote access to higher end facilities that might contribute to the K–12 education process. (Part 3: Resources)
   - **Finding:** Person-to-person contact remains the most effective approach to education.
   - **Recommendation:** The various university-based NanoCenters mobilize their undergraduate and graduate students to engage in K–12 education at the nanoscale. Federal funding agencies must provide an adequate budget allowance for this work. Universities

must recognize the faculty supervisory efforts in tenure and promotion decisions.

- **Finding:** To regain prominence in science, technology, and engineering the U.S. must stop having a haphazard approach to curriculum development.
- **Recommendation:** Funding is needed to allow for the design, development, testing, and implementation of a coherent curriculum that would allow 7- to 16-year-old students to develop an integrated understanding of core science ideas that underpin nanoscience and engineering. Such a curriculum would focus on helping students develop progressively deeper understanding of core ideas. Such a process calls for change in the standards that focus on teaching big ideas with a focus on developing a deeper understanding of these ideas.
- **Principal Stakeholders include**: NSF, DoEd, NSTA, and professional science and engineering societies.

5. Informal Education through Museums:
   - **Finding:** To date, the Nanoscale Informal Science Education Network (NISE Net) has largely examined how "nano" fits into the size scale of materials. Several studies have suggested that students will respond best to STEM as it addresses societal problems. Now that nano-enabled technologies are beginning to proliferate, it would be timely to develop exhibits and programs associated with the impact of those nano-enabled technologies.
   - **Recommendation:** NSF, which is the principal funding source for new nanoscale science and engineering projects, should take the lead in establishing links between museums and the national and international research communities for new exhibit development. Moreover, other Federal funding agencies and industry representatives must also be contributors since they will be engaged in the translational efforts that lead to technology impact.
   - **Principal Stakeholders include:** NSF, other relevant mission-oriented Federal funding agencies, Museums, ASTC, and NanoBusiness Alliance.
6. Public Education through Media:
   - **Finding:** Nano-enabled technologies will enable us to alter our world and provide advances in standards of living. With the decline in the number of science journalists, there is an opportunity for the NNCO, University and Industrial programs, and other stakeholder groups to develop a continuing stream of information that can inform the public of the benefits and risks

emanating from progress at the nanoscale. The rapid growth in information technologies is creating new interaction paradigms that might be exploited using electronic media (e.g., Wikipedia, Facebook, Second Life, YouTube, and Kindle) that are being adopted by young and IT-literate learners.

- **Recommendation 1**: Cyber-education should be included in the suite of learning venues to engage students. NSF, with its interest in cyberlearning, should take the initiative but the DoEd must be engaged to ensure a continuing effort.
- **Recommendation 2**: The Wikipedia entries on nanotechnology should be routinely updated and expanded. This task might be best accomplished by mobilizing the variety of talent and expertise at the various NanoCenters. K–12 science teachers should be involved to ensure the information is structured in ways that can be readily absorbed at the various grade levels.
- **Principal Stakeholders include:** NSF, DoEd and other agencies with relevant missions, NSET/NNCO, NSTA, and ASTC.

7. Local Community Outreach and Engagement:
    - **Finding:** The national media plays an important role in informing people. However, local and personal engagement is often more effective.
    - **Recommendation 1**: Existing NanoCenters should expand their outreach activities to local and state communities.
    - **Recommendation 2:** A Nanoscale Science and Engineering Education (NSEE) forum should be established to share best practices wherein all federally funded NanoCenters should participate, not just those funded by NSF.
    - **Principal Stakeholders:** NSET/NNCO, ASTC, and NGA.
8. Universities and Community Colleges:
    - **Finding:** While the U.S. research universities are acknowledged as world leaders in science and engineering, there is growing global competition—especially in nanoscale science, engineering, and technology. To meet the national needs in the near future, it will become more important to foster U.S.-born students in STEM. Since 40% of college students get their start in community colleges, the Opportunity Equation report[10] recommends closer interaction between community colleges and the universities.
    - **Recommendation:** NSF, DoEd, and other agencies with relevant missions should foster nanotechnology curricula development and evaluation that is appropriate for community colleges and ensure meaningful collaborations between the community

colleges and the NanoCenters. The DoEd's Department of College and Career Transitions program addressing articulation should ensure nanotechnology is included in that program.

- **Principal Stakeholders:** NSF, DoEd, Department of Labor (DoL), universities, and National Association of Community College Entrepreneurship (NACCE).

9. Industrial Needs:
   - **Finding:** Preparation for employment is an important aspect of the educational process. In our rapidly evolving world, the needs of industry are fluid due to changing technologies and growing global competition. Nanoscale science will be instrumental in technological change. Many countries have followed the U.S. lead and established a nanotechnology initiative. Moreover, those initiatives tend to be more focused on targeted technology development than is the U.S. Consequently, there will likely be strong competition for nano-trained people between the U.S. and those countries.
   - **Recommendation:** The DoL needs to work with industry groups and professional science and engineering societies to develop accurate assessments of domestic workforce needs, including the effects of growing overseas education and job opportunities.
   - **Principal Stakeholders include:** DoL, Department of Commerce (DoC), professional science and engineering societies, and NanoBusiness Alliance.

10. Cyber and Virtual Innovations:
    - **Finding:** NSF's investments in cyber-infrastructure along with those of other agencies have resulted in a range of state-of-the-art, distributed digital resources: some for nanotechnology and science research, some for nanotechnology learning and education, and some for nanotechnology events and news. These cyber-nanotechnology resources vary in terms of their target audiences (education level, country of origin, original purpose), quality, level of integration with other cyber-infrastructure resources, and usage and usage reporting. Most lack such meta-information, which may limit knowledge transfer across user communities in terms of discoverability, searchability, and adaptability. The emerging NanoEducation community must be able to exploit existing cyber-infrastructure resource investments more effectively.
    - **Recommendation 1**: NanoHUB, National Center for Learning and Teaching (NCLT), NNIN, and other cyberinfrastructure resources focused on nanotechnology need to be better publicized regarding accessibility, targeted user-levels, customizability both

in terms of targeted audiences and user interface, interoperability with other systems, and service and training offerings.

- **Recommendation 2:** Consideration should be given to the research and development of an overall mechanism for efficient search, access, and use of cyber-infrastructure resources focused on nanoscience and technology with potential relevance to education at all levels. Such a mechanism would likely entail providing for the creation of an inventoried, analyzed, tagged registry. Also worth considering are mechanisms for enabling greater interoperability with emerging cyber-nanotechnology resources such as remote access and control to state-of-the-art nanotechnology characterization equipment; databases for properties, applications; environmental, health, and safety implications; and general educational paradigms, such as virtual and immersive environments, simulations, and games. Such integration and knowledge sharing offers greater promise for accelerated discovery and learning of nanoscience and technology.
- **Principal Stakeholders include:** NSF, other Federal agencies with relevant missions, NSET/NNCO, NSTA, Open Education Resources (OER), and The NanoTechnology Group Inc.

Changes to No Child Left Behind by President Obama in Proposed 2011 Budget

While I was writing this chapter, announcements were made by President Obama that may prove to be helpful in developing solutions to the various issues under discussion.

> President Barack Obama announced that his administration will send to Congress the blueprint for an updated Elementary and Secondary Education Act that will overhaul No Child Left Behind. The plan will set the ambitious goal of ensuring that all students graduate from high school prepared for college and a career, and it will provide states, districts and schools with the flexibility and resources to reach that goal.[11]

Preparing Students for Success in College and the Workforce

> We will end what has become a race to the bottom in our schools and instead spur a race to the top by encouraging better standards and

> assessments... And I'm calling on our nation's governors and state education chiefs to develop standards and assessments that don't simply measure whether students can fill in a bubble on a test, but whether they possess 21st century skills like problem-solving and critical thinking and entrepreneurship and creativity. That is what we'll help them do later this year—when we finally make No Child Left Behind live up to its name by ensuring not only that teachers and principals get the funding that they need, but that the money is tied to results.
>
> **President Barack Obama**
> *Remarks to the Hispanic Chamber of Commerce, March 10, 2009*

Last year, the President challenged states to develop standards and assessments to help America's children rise to the challenge of graduating from high school prepared for college and the workplace.

A year later, the President applauded governors for their efforts to work together in a state-led consortium—managed by the National Governors Association (NGA) and the Council of Chief State School Officers (CCSSO). They were tasked to develop and implement new reading and math standards that build toward college- and career-readiness. Many states are now well positioned to adopt these standards, proving the governors' initiative was an essential first step in improving the rigor of teaching and learning in America's classrooms.

Are the States Cooperating?

Since last year, 48 states have been collaborating to write common standards in math and reading, coordinated by the governors' group with the encouragement of the White House.

According to an article[12] in the *New York Times,* Texas and Alaska decided not to participate in that state-led effort. Unless the new standards become mandatory under planned revisions of No Child Left Behind, this decision could affect 46 or 47 states that must select textbooks based on the current Texas standards.

The common-standards effort has produced a draft, and earlier this month Kentucky became the first state to approve the substitution of the new standards for the state's own standards in the two subjects.

How successful or quickly the governors, legislatures, state boards of education, and other authorities in other states will be in agreeing on adoption of the new standards is not clear.

"In better aligning the law to support college- and career-ready standards," the White House statement said, its proposed rewrite of the No Child Left Behind law would "require all states to adopt and certify that they have college- and career-ready standards which may include common standards developed by a state-led consortium, as a condition of qualifying for Title I funding."

"In its 2011 budget request, the White House also hoped to replace the law's much-criticized school rating system, known as adequate yearly progress, with a new accountability system."

Raising the Rigor of Academic Standards

Economic progress and educational achievement go hand in hand. The national imperative to educate every American student to graduate prepared for college and for success in a new workforce cannot be ignored. In 1994, the Elementary and Secondary Education Act (ESEA) established a requirement that each state set standards for what America's students should know and be able to do in critical subjects. But the law did not ask states to consider whether those standards were aligned with what is needed for success in college and in the workplace.

According to the National Center for Education Statistics, 31 states set proficiency standards for fourth-grade reading lower than even the basic level measured by the National Assessment of Educational Progress (NAEP). Under the No Child Left Behind Act, between 2005 and 2007, various states have also lowered their standards in reading and math.

This race to the bottom has placed American students on a decline in relation to international peers. Results on international assessments reveal that, in math, American students lag almost a full year behind students from the top-performing countries. In response to their international comparison results, other countries have raised their standards, while ours have been lowered.

President Obama has encouraged the state-led effort to end the practice of lowering state reading and math standards, and encouraged developing initiatives that support states in the adoption and implementation of college and career-ready standards. To spur reform and excellence in our schools, the President's Race to the Top competition rewards states that choose to adopt state-developed common standards that build toward college- and career-readiness. In the coming months, the Obama Administration will also commit $350 million to a new competition supporting state-led partnerships to develop new, state-of-the art assessments aligned to college and career-ready standards. In the past decade, the states have had to pay various testing companies from their own state budgets to develop the various mandated tests required by federal law.

In better aligning the law to support college- and career-ready standards, the Obama Administration will integrate new policies into a redesigned Elementary and Secondary Education Act, which will:

- Require all states to adopt and certify that they have college- and career-ready standards in reading and mathematics, which may include common standards developed by a state-led consortium, as a condition of qualifying for Title I funding.
- Include new funding priorities for states with college- and career-ready standards in place, as they compete for federal funds to improve teaching and learning and upgrade curriculum in reading and math. This priority applies to the President's FY2011 budget request for new Effective Teaching and Learning programs in literacy ($450 million) and STEM ($300 million).
- Encourage states, schools districts, and other institutions to better align teacher preparation practices and programs to teaching of college and career-ready standards. This priority supports the President's FY2011 budget request for a new Teacher and Leaders Pathways program ($405 million).
- Assist states in implementing assessments aligned with college- and career-ready standards, under a new Assessing Achievement program. The President's FY2011 budget supports $400 million in state grants under this program.
- Support the expansion of the Race to the Top, beyond funding in the Recovery Act, to dedicate $1.35 billion in awards to states and school districts that have college- and career-ready standards in place as a condition of funding.
- Support professional development for teachers, leaders, and other school instructional staff to better align instruction to college and career-ready standards. This supports the President's FY2011 budget request for the Effective Teacher and Leaders state grant program ($2.5 billion).

The *Educate to Innovate Program*[13] Addresses Science and Technology—STEM Education

President Obama has launched an *"Educate to Innovate"* campaign to improve the participation and performance of America's students in science, technology, engineering, and mathematics (STEM). This campaign will include efforts not only from the Federal Government but also from leading companies, foundations, nonprofits, and science and engineering societies to work with young people across America to excel in science and math.

As part of the campaign, this Administration hopes to do a series of events, announcements, and other activities that build upon the President's "call to action" and address the key components of national priority.

Why Is This Important?

We have many great schools, excellent teachers, and successful students in America. But there are also troubling signs that, overall, our students should be doing better in math and science.

- In the *2006 Program for International Student Assessment (PISA)*[14] comparison, American students ranked 21st out of 30 in science literacy among students from developed countries, and 25th out of 30 in math literacy.
- On the *2009 National Assessment of Educational Progress (NAEP)*[15] math tests, 4th graders showed no signs of progress for the first time in many years, and 8th graders tallied only modest evidence of progress. We are not advancing as we must.

What We Must Do

Through *"Educate to Innovate"* and other efforts, we must:

- Increase STEM literacy so that all students can learn deeply and think critically in science, math, engineering, and technology.
- Move American students from the middle of the pack to top in the next decade.
- Expand STEM education and career opportunities for underrepresented groups, including women and girls.

The First Steps

America is already stepping forward to meet these challenges. As part of the *"Educate to Innovate"* effort, five major public–private partnerships are harnessing the power of media, interactive games, hands-on learning, and community volunteers to reach millions of students over the next four years, inspiring them to be the next generation of inventors and innovators.

- Time-Warner Cable,[16] Discovery Communications,[17] Sesame Street,[18] and other partners will get the message to kids and students about the wonder of invention and discovery.
- National Lab Day[19] will help build communities of support around teachers across the country, culminating in a day of civic participation.

- National STEM design competitions[20] will develop game options to engage kids in scientific inquiry and challenging designs.

Five leading business and thought leaders (Sally Ride, Craig Barrett, Ursula Burns, Glen Britt, and Antonio Perez) will head an effort to increase private and philanthropic involvement in support of STEM teaching and learning.

How Do We Start the Implementation of Resources?

The government has funded an enormous amount of resources over the past decade to bring science education into the twenty-first century matrix. While Congress addresses the next phase of changes in the No Child Left Behind Act, many of these programs can be implemented into our schools. The next chapter will bring some of these ideas to the forefront for consideration in your school district.

New Legislation That Addresses Workforce Training of Technicians in Nanotechnology

Senate Bill S 3117 IS—Promote Nanotechnology in Schools Act[21] (Introduced in Senate)

To strengthen the capacity of eligible institutions to provide instruction in nanotechnology, this bill will strengthen the ability to introduce nanoscience in secondary education, along with 2- and 4-year colleges and universities. The following excerpts from the bill define the intent.

(a) Findings—Congress makes the following findings:

(1) The rapidly growing field of nanotechnology is generating scientific and technological breakthroughs that will benefit society by improving the way many things are designed and made.

(2) Nanotechnology is likely to have a significant, positive impact on the security, economic well-being, and health of Americans as fields related to nanotechnology expand.

(3) In order to maximize the benefits of nanotechnology to individuals in the United States, the United States must maintain world leadership in the field, including nanoscience and microtechnology, in the face of determined competition from other nations.

(4) According to the National Science Foundation, foreign students on temporary visas earned 33 percent of all science and engineering doctorates awarded in the United States in 2007, the last

year for which data are available. Foreign students earned 63 percent of the engineering doctorates.

(5) To maintain world leadership in nanotechnology, the United States must make a long-term investment in educating United States students in secondary schools and institutions of higher education, so that the students are able to conduct nanoscience research and develop and commercialize nanotechnology applications.

(6) Preparing United States students for careers in nanotechnology, including nanoscience, requires that the students have access to the necessary scientific tools, including instruments designed for teaching, and requires training to enable teachers and professors to use those tools in the classroom and the laboratory.

(b) Purpose—The purpose of this section is to strengthen the capacity of United States secondary schools and institutions of higher education to prepare students for careers in nanotechnology by providing grants to those schools and institutions to provide the tools necessary for such preparation.

(c) Authorization of Appropriations—There are authorized to be appropriated to the Director to carry out this section $15,000,000 for fiscal year 2010, and such sums as may be necessary for each of the fiscal years 2011 through 2013.

Latest Major Action: 3/15/2010 Referred to Senate committee. Status: Read twice and referred to the Committee on Health, Education, Labor, and Pensions.

Current Workforce Training Programs

The five states (Pennsylvania, New York, Minnesota, Texas, California) that have addressed workforce programs for nanotechnology technicians can serve as models for other states who may receive proposed funding if this Bill is passed. *Review Part 3: Chapter 10* for current programs.

A Conceptual Framework to Develop New Science Education Standards for K–12

In order to develop a new STEM EcoSystem of Education that includes nanoscale science, we must look at the work that was recently done by the National Research Council's (NRC) Committee on "A Conceptual Framework to Develop New Science Education Standards for K–12."[22]

A draft of the framework was available for public comment during the period from July 12 through August 2, 2010. Over 2,000 people responded to the online survey and hundreds more participated in discussion groups across the country. The committee received a wide range of comments from a wide variety of perspectives. This was the first opportunity to address the inclusion of nanoscale science as a size for scientific inquiry in the new science education standards for K–12.

The NRC committee has begun the process of summarizing and digesting the wide variety of suggestions expressed through the feedback and will be deciding on appropriate revisions to the framework as well as finishing writing the rest of its report. This process will take several months. Once the revisions are complete, the framework will undergo the traditional National Research Council confidential review by a diverse group of experts. After the committee revises its report in response to reviewers' comments, the framework report will be finalized and released to the public in early 2011.

The nonprofit education organization Achieve (www.achieve.org) will then work with a group of states to develop a set of standards for K–12 science education based on and guided by the final NRC committee framework report. Achieve's science work will be led by Stephen Pruitt, who has recently joined the Achieve staff (having resigned as a member of the NRC framework committee for this appointment). Achieve has already begun planning and is currently developing a network of state partners. Further opportunities for public comment will be managed by Achieve as its work on the science education standards proceeds in 2011.

References*

1. http://www.nnin.org/nnin_educstandards.html.
2. NSF 0654-5 Appendix2, A Nanoleap—K–12 & Informal Nanoscale Science and Engineering Education (NSEE) in the U.S.
3. http://www.nnin.org/nnin_educstandards.html.
4. http://www.project2061.org/events/workshops/default.htm.
5. http://www.project2061.org/publications/bsl/online/index.php.
6. http://strandmaps.nsdl.org/.
7. http://www.project2061.org/publications/textbook/mgsci/report/mgbooks.htm.
8. NSF award EEC 0805207 Partnership for Nanotechnology Education, http://www.nsf.gov/crssprgm/nano/reports/nsfnnireports.jsp.
9. http://www.achieve.org/K–12 Benchmarks.

* All links active as of August 2010

10. The Opportunity Equation: Transforming Mathematics and Science Education for Citizenship and the Global Economy, Carnegie Corporation of New York and Institute for Advanced Study, 2009, http://www.OpportunityEquation.org.
11. http://www.whitehouse.gov/the-press-office/weekly-address-president-obama-send-updated-elementary-and-secondary-education-act-.
12. http://www.nytimes.com/2010/02/22/education/22educ.html?th&emc=th.
13. http://www.whitehouse.gov/issues/education/educate-innovate.
14. http://www.oecd.org/document/2/0,3343,en_32252351_32236191_39718850_1_1_1_1,00.html.
15. http://nces.ed.gov/nationsreportcard/mathematics/.
16. http://connectamillionminds.com/.
17. http://science.discovery.com/fansites/be-the-future/be-the-future.html.
18. http://www.sesameworkshop.org/home.
19. http://www.nationallabday.org/.
20. http://www.dmlcompetition.net/.
21. http://thomas.loc.gov/cgi-bin/query/z?c111:S.3117:
22. Helen R. Quinn, PhD (NAS), Chair, NRC Committee to Develop a Conceptual Framework for New Science Education Standards, Chair, NRC Board on Science Education.

14

The Twenty-First Century Paradigm—Working Together

> There is plenty of room at the bottom.
>
> **Richard P. Feynman**

The education paradigm and nanoscale science have a lot in common. Now that you have reviewed the resources and government support "from the top down," it is time to discuss the potential for teachers to explore the materials and start teaching "from the bottom up."

Where Do We Start?

Maybe teachers could start with familiar technology. Introducing nanoscience potential with this attention-grabbing video of a shape-shifting cell phone like *Morph* may pique teachers' and students' interest in learning about nano-enabled advanced technology.

Nokia *Morph* Concept Video[1]

Nano concepts can be introduced to students with this YouTube Video that shows a nanotechnology-designed cell phone. *Morph* is a concept demonstrating some of the possibilities nanotechnologies might enable in future communication devices. *Morph* can sense its environment, is energy harvesting, and self-cleaning. *Morph* is a flexible two-piece device that can adapt its shape to different use modes. Nanotechnology enables adaptive materials that can morph into rigid forms on demand.

It is also featured in the Museum of Modern Art's online exhibition, *Design and the Elastic Mind.*[2] It has been a collaboration project of the Nokia Research Center and Cambridge Nanoscience Center.

Teachers and Students Can Explore the Curriculum and Resources Together

Teachers and students can start the process by challenging each other to find a new resource listed in Part 3: Chapter 9, and explore nanoscale science each week as "inquiry," then discuss how the resulting technology will affect their future. The curriculum developed as K–12 Outreach has already been assessed, and Teachers' Guides are provided for most lessons.

Whoever finds the lesson makes a presentation to the class, stimulates the discussion about how it relates to their world, and conducts an inquiry at one of the interactive Internet sites. *Ask the Scientist*[3] connects your students to some of the top scientists in the country specializing in bio-nanosciences, with connections to the Howard Hughes Medical Institute. They can answer questions about medicine, human biology, animals, biochemistry, microbiology, genetics, or evolution.

What About Physics, Chemistry, or Engineering?

If you have questions on other branches of nanoscience visit the Argonne National Laboratory–NEWTON BBS.[4] BBS has posted over 20,000 questions since its inception in 1991, which are archived on the site. This unique service is intended for K–12 teachers and students. The question submitted is e-mailed directly to the volunteer scientists globally, and answers are e-mailed to the return e-mail address provided. Questions and answers are also posted within NEWTON BBS and are archived for others to access in the future. Teachers and students in the elementary grades can also learn together with some of the Informal Science resources listed in Part 3: Chapter 11.

New Programs to Support Teachers

Many new programs are introduced on a regular basis to help teachers change their classrooms with technology. The following resources and programs are recent additions that can be helpful.

National Lab Day[5]

National Lab Day is more than just a day. It is a nationwide initiative to build local communities of support that will foster ongoing collaborations among volunteers, students, and educators.

Volunteers, university students, scientists, engineers, other STEM professionals, and more broadly, members of the community are working together with educators and students to bring discovery-based science experiences to students in grades K–12 . When an educator posts a project, the system will help them get the resources needed to bring that project to fruition.

The National Lab Day Initiative currently have projects throughout the country—check the Interactive Map[6] to review your area.

Curriki Curriculum Development Site for Teachers[7]

Curriki was developed for the community of educators, parents and students. The developers started a forum for open source nanoscience curriculum and work together to develop interesting, creative, and effective educational materials that the global educational community can use for free. Their goal is to develop curriculum through community contributors, deliver the curriculum globally, and determine the impact by project and by individual. The initial focus is on K–12 curricula in the areas of mathematics, science, technology, reading and language arts, and languages. As I made my inquiries, there was definitive interest in having the teachers start a nanoscience forum for K–12 teachers.

Check out the Focus on Science Section.[8]

Desktop Scanning Tunneling Microscope (STM) and Atomic Force Microscope (AFM) Scanners for Nanoscience Education[9]

STM and AFM desktop scanners from NanoSurf provides ease of use, mobility, and affordable pricing allows schools to easily set up teaching labs and outreach programs, nanoscience curriculum, books, software, animated gallery, teacher demos.

The Virtual Nanolab at University of Virginia was designed with these tools for developers. If your school receives a technology grant for science, a desktop STM would be the perfect instrument for a high school science lab in the district that could also accommodate elementary students with nano-lab visits. As high school students learn to use the equipment, they can volunteer to demonstrate and help younger students learn.

TeachNano[10]

TeachNano seminars provide a unique opportunity for those involved in teaching nanoscience to learn about other nanoscience educational programs and to share experiences. The program includes short talks, hands-on demonstrations, and open discussions.

Project Share: Designed for Texas Teachers in 2010 and Students in 2011

The Texas Computer Education Association (TCEA) announced in the spring of 2010 that *Project Share*[11] will be available for teachers. The Texas Education Agency Initiative entitled *Project Share* is a collaboration with Epsilen LLC, and *The New York Times Company Knowledge Network*,[12] developed to help teachers work in collaborative environments, develop e-portfolios, and share professional learning communities.

Engaging Learning around Shared Digital Content

The Texas Education Agency (TEA) Initiative supports professional teachers' development in an interactive and engaging learning environment. The single online platform leverages existing and new resources for teachers across the state and supports learning communities where educators can collaborate, create, and share. Teachers can share the K–12 nanoscience resources among their peers with the new Project Share.

Groups of teachers across Texas will be able to meet together online, establish a central repository for professional learning materials, and use Web 2.0 collaboration tools such as wikis, blogs, drop boxes, and chat rooms. The system provides online lesson plan tools, class e-mail and calendars, and peer review that leverages a multimedia approach crucial for today's teachers. Project Share offers the ability for educators to plan, manage, and tailor professional development based on individual needs.

Teachers and students have access to the *New York Times* repository, which will allow teachers to supplement lesson plans and assignments with credible sources and citations. For example, a science teacher building a unit on refraction can in minutes pull video and audio files from NASA, National Science Foundation, and higher education experts in the field. The Epsilen network and *New York Times* repository allow teachers to leverage existing content and specifically align it with Texas Essential Knowledge and Skills.

Texas Virtual School Network

The system will allow teachers to develop online courses and will work in conjunction with the *Texas Virtual School Network*[13] (TxVSN). TxVSN was authorized by the Texas Legislature in 2007 to provide additional opportunities and options for Texas students through online courses.

Project Share also extends beyond public school teachers and students. It will be available for all local education entities, including charter and private schools. There will be some participation from local colleges and universities as well.

Bring Your Own Technology (BYOTech)

An experimental project that allows students to bring their own technology to schools is taking place in some states. Most students have their own laptops, iPhones, iPod touches, and other computing tools that can connect to the district networks. This "Bring Your Own Technology" approach is slowly gaining favor across the country as school districts struggle with tight budgets and the need for greater technology access for students and staff members.

The program involves an increase in the schools' network capacity, along with its bandwidth and the addition of wireless access points. Teachers and administration officials would need to collaborate to discuss the change in policy and the implementation. Meetings would also need to be held for parents to discuss the changes to policy.

Questions to consider as you look at these solutions:

What are the possibilities of BYOTech in your school district?

What problems might this help to solve?

What new problems might be created?

Laptops for Public Schools

Many states are already providing laptops for high school students, while others are including laptops for K–12 grades as interactive virtual schools are changing classrooms.

eSchool—An Excellent Resource for Teachers of K–20

Innovation in Education: Aligning Teacher Effectiveness to Greater Student Achievement

The most important lever in student achievement is the effectiveness of teachers, administrators, and other support staff. Knowing how to "adjust that lever" can be a challenge. School district leaders today are asking themselves "What does an effective teacher look like?" or "What processes can I put in place to motivate and retain the best and the brightest?" Webinars[14] are scheduled regularly for 1 hour each plus 15 minutes of questions and answers.

A variety of webinars on the hottest topics in education technology are listed on the eSchool news Web site. The webinars are a great place to learn about new technologies, how your colleagues are solving concerns you may have, and more. All you will need on the day of the event is Internet access. After you register, you will be supplied with a link to log-in at the scheduled time.

The webinars are also recorded and archived. Teachers can go back and review a webinar presentation after the event and share it with colleagues.

Teachers' Online Resources for Twenty-First Century Learning

The Internet site also has abundant resources for teachers, a repository of videos with up-to-date information and technology solutions. The Web site has three sections for you to explore:

eSchoolnews.TV[15]

Connects you to the latest videos on technology news and information resources including video interviews, marketplace news, and student videos. Over 400 videos archived.

eSchool Classroom News[16]

This section keeps you up to date with the latest education news headlines and has links to the resource centers.

Educator Resource Centers[17]

- One-to-One Computing: The Last Piece of the Puzzle
- Creating the 21st-Century Classroom
- Securing Your Schools from the Inside Out
- Next Generation Collaboration
- Igniting and Sustaining STEM Education
- Securing Student Laptops for Safe Learning
- Effective Programs for Dealing with Autism in Schools Today
- Measuring 21st-Century Skills
- Successful Video Production

eSchool Campus News[18]

This section keeps you up-to-date on all educational news for colleges and universities. Sections for conferences, curriculum, and funding.

Alternative Solution for High Schools

Insight Virtual High Schools[19]

Insight Schools operates a national network of accredited, diploma-granting, tuition-free online high schools. The schools offer highly personalized education, designed to accommodate individual learning styles, lifestyles and goals, while preparing students for the twenty-first century workplace. Also an alternative solution for students who consider dropping out of high school.

Unique Challenges

Teachers, parents, and students are facing many unique challenges as the twenty-first century technological race moves at warp speed. The teacher development tools and resources are intended to support your combined efforts to introduce nanoscience K–12 education in your classrooms.

References*

1. http://www.youtube.com/watch?v=IX-gTobCJHs&feature=player_embedded
2. http://www.moma.org/interactives/exhibitions/2008/elasticmind/
3. http://www.askascientist.org
4. http://www.newton.dep.anl.gov/aas.htm
5. http://www.nationallabday.org/
6. http://www.nationallabday.org/projects_map
7. http://www.curriki.org/xwiki/bin/view/Main/
8. http://www.curriki.org/xwiki/bin/view/Main/FocusOn_Science
9. http://www.nanoscience.com/education
10. http://www.teachnano.com/teach/
11. http://projectsharetexas.org
12. http://www.nytimes.com/college
13. http://www.txvsn.org/
14. http://www.eschoolnews.com/events/webinars
15. http://www.eschoolnews.tv
16. http://www.eschoolnews.com/publications/eclassroom-news/
17. http://www.eschoolnews.com/resources/educator-resource-center/
18. http://www.ecampusnews.com/
19. http://www.insightschools.net/

* All links active as of August 2010.

15

Rethinking Education—Can We Succeed?

> I never teach my pupils. I only attempt to provide the conditions in which they can learn.
>
> **Albert Einstein**
> *U.S. (German-born) physicist (1879–1955)*

This section is dedicated to teachers and the profession of teaching. Many teachers are rethinking the present educational modalities and are electing to enlist students into the process of reshaping classroom instruction.

If you remember Figure 2.1 in Chapter 2, *The Curriculum Communities of Stakeholders*, you will notice that the students were missing. They are the learners and a very important component of the matrix for implementation. Since teachers are the direct line of communication for their learning experience, both need the total support of all levels of stakeholders for education to improve.

Inclusion of Students for Solutions

The question arises of how to best support teachers in this rapidly evolving education landscape, which involves so much new technology. Some are in favor of retraining teachers, and others believe in allowing students and teachers to develop their own solutions. The answer may be that both are viable and productive solutions to consider. Soliciting student input also encourages more involvement and responsibility and stimulates a vested interest in learning.

Consider this statement by President Obama, which clearly defines the issues facing teachers:

> Education, then, is what has always allowed us to meet the challenges of a changing world. And that has never been more true than it is today. You're coming of age in a 24/7 media environment that bombards us with all kinds of content and exposes us to all kinds of arguments, some of which don't rank all that high on the truth meter. With iPods and iPads, Xboxes and PlayStations, information becomes a distraction, a diversion, a form of entertainment, rather than a tool of empowerment. All of this is not only putting new pressures on you; it is putting new pressures on our country and on our democracy. It's a period of breathtaking change,

like few others in our history. We can't stop these changes, but we can adapt to them. And education is what can allow us to do so.

President Obama
Commencement address[1] at Hampton University on May 9, 2010

Re-Inventing Schools Coalition

Schools in Alaska were the first to try this new "Re-Inventing Schools"[2] program with success. Colorado and Maine have implemented the program in primary grades in some of their districts.

The organization has a shared vision for communities of stakeholders to drive systemic changes in their districts. In this vision, the education community speaks as one voice about the skills and knowledge that all students should acquire, which is the foundation of the program. Their voices are also respected regarding the educational values and how those would be realized over time. The shared vision process has great value as the community is deliberately drawn into the system and any walls between school and the community come down. I have included a sample of their leading questions that stakeholders can use to start the process of discussion in their districts or with the organization.

Sample leading questions for stakeholders (staff, student, board, and community members) to create a shared vision:

- What do we want to keep that we are doing well?
- According to our current results, how are our students doing?
- What do our students need to know and be able to do in the twenty-first century?
- If needed, how should we change the current system?

"What Works Clearinghouse"[3] is an Internet site developed by the U.S. Department of Education that provides research based on "what works" for teachers and administrators. The information is sparse, but one research report about a school in Harlem has been prominently advertised nationwide in a TV commercial.

After viewing the television advertisement about the Promise Academy Middle School and the Harlem Children's Zone®, it was interesting to find the original reports on the What Works Clearinghouse. This is an example of a "bottom-up" approach involving community stakeholders that features improved outcomes.

The following is a quick review of the article "Are High-Quality Schools Enough to Close the Achievement Gap? Evidence from a Social Experiment in Harlem"[4] which features the Promise Academy Middle School–Charter school sponsored by the Harlem Children's Zone.

These were the first changes made to implement their programs:

- Extended school day and year with additional after-school tutoring and Saturday classes
- Intensive test preparation including morning, mid-day, after-school, and Saturday sessions
- Student incentives for high achievement, such as money and trips to France
- School health clinic provides students free medical, dental, and mental-health services

What Was the Study About?

This study examined the effects on academic achievement of offering students enrollment in the Promise Academy charter middle school. The school is sponsored by the Harlem Children's Zone®, which combines reform-minded charter schools with a web of community services designed to provide a positive and supportive social environment outside of school.

The study analyzed data on about 470 New York City students who applied for enrollment in 2005 and 2006 as entering sixth graders. The number of applicants exceeded the school's capacity, so enrollment offers were granted by random lottery.

The study measured effects by comparing the outcomes of students who were selected in the lottery and offered enrollment in the school to students who were not selected in the lottery.

Student outcomes were measured in sixth, seventh, and eighth grades using standardized statewide math and English language arts (ELA) tests.

What Were the Outcomes?

Students offered enrollment in the school had higher math test scores in sixth, seventh, and eighth grades than the students not offered enrollment. By the time they were tested in eighth grade, the effect size for the math test was 0.55. The WWC interprets this as equivalent to moving a student from the 50th to the 71st percentile.

The study authors found no statistically significant differences in ELA test scores in sixth or seventh grade, but a positive effect was found on the eighth grade ELA test. The effect size was 0.19, which the WWC interprets as equivalent to moving a student from the 50th to the 58th percentile.

These examples show that stakeholders in a community make a positive difference. The school is not currently teaching nanoscience in their classrooms. However, it is a good example that proves operating as a community ensures stakeholders an opportunity to review new educational resources to make informed decisions.

Teachers Respected As Stakeholders

The final story involves a project for an entire country based on collaboration between professors and "seed teachers" to design exciting innovative nanoscience curriculum for K–12 students. I was honored to escort the group of six professors and two teachers from Taiwan in 2005, to share their story with Arizona State University and University of Wisconsin–Madison, MRSEC. The results of their project from 2002 to 2007 are presented as a model for any country that wishes to succeed in developing nanoscience education for K–12.

"Seed Teachers" Are the Key to Developing Nanoscience Curriculum

The inclusion of "seed teachers" from the beginning of the project development was the key to the success of the program. The teachers learned the concepts from the lectures at the universities, then developed curriculum based on their understanding. The Taiwan Nanotechnology Initiative government officials realized that not all students would become scientists, but everyone needed to be educated to understand the basic concepts of nanotechnology. Realizing that teachers are not scientists either, they planned focused lectures on basic concepts from the integrated subjects. Teachers took notes to develop lessons for textbooks that students could relate to, along with videotaped experiments. The animations, comic books, and games they consulted on were designed for the younger students that were already "digital natives."

Obstacles Still Prevail in the United States

The United States has developed many resources—provided in Part 3—without inclusion of the teachers. The cognitive scientists are still studying how children learn, and the universities that created the outreach materials are concerned that teacher training to use the curriculum may be necessary. The few teachers that have been trained at the universities do not know where to place the new curriculum in their syllabus. That important step was not considered in any of the grants to develop the outreach programs, because nanoscience has not been specified in the national or state standards.

Teachers Teach the Teachers

It is not too late for our teachers to get started. The universities that developed the outreach programs could start the process by enlisting science teachers as "seed teachers," with an elective course on team teaching. Each "seed teacher" would be expected to teach classes in their district upon completion. This would multiply exponentially as the "seed teachers" explained the program in their home schools and enlisted teachers to sign up for their courses.

Students could also get involved as *Nano Ambassadors* to align fellow students in the quest to learn more about nanoscience and the emerging technologies, much like the student in Chapter 3, who enlisted the principal of the school, when he wanted his school to "go green" with environmental education. If the teachers and the students both display a passion for learning, the administration will join the effort. Based on all the activity globally, it is apparent that we do not have another decade to wait for textbooks and new standards to be approved. The world is not waiting, and neither should our students.

The following information is based on my report from the unique collaboration in 2005, and the most recent report of further development of the K–12 instructional materials in Taiwan from 2005 to 2007.

The First International Collaboration in the United States on K–12 Nanoscience Courses[5]

A long-held dream of a team of engineering professors and teachers from Taiwan to visit the United States and share their project for K–12 nanoscience education became a reality in 2005. Taiwan has a system very much like the United States in that they do not have a national curriculum mandate. This allowed them to "think out of the box" and devise a method that included teachers from the beginning. Their program is titled *Nanotechnology Human Resources Development (NHRD),*[6] which was organized to address the importance of having skilled and dedicated workers for their technological advancement in the emerging integrated fields of science. The amount of resources the teachers created since 2002 should serve as an inspiration to other teachers. Developing an array of textbooks, animated videos, comic books, coloring books, and educational video games in a short span of time, all based on their understanding of the university lectures concerning nanoscale science, proves that specialized training for teachers is not necessary. It is my hope that our teachers will garner the same respect from the universities which have developed curriculum without their input. Closing the gap in the communities of curriculum development—so that the "top-down" stakeholders learn to work with the teachers and students "from the bottom up" as stakeholders—would bring excitement into the process.

The North Region K–12 Education Center for Nanotechnology

The twenty-first century is considered to be a knowledge-based economic society with advanced technology. Nanotechnology has become a most important science and technology for the world. Since 2002, the Executive Yuan of Taiwan has conducted a nanotechnology national project. Some funds from this national project have been devoted to human resources development. Further, to fulfill the needs of human resources for the next ten, twenty, even thirty years in the future, the education for the present K–12 young students is essential. The program emphasizes nanotechnology educational programs at the K–12 level. This provides teachers information to develop material to inspire students to learn about advanced technology. In cooperation with the Ministry of Education, National Taiwan University was invited to establish the one of the five regional centers in Taiwan. The North Regional Center is the leader to form a union with K–12 schools from the four other regional centers. In the first year, the common knowledge of nanotechnology has expanded to the off-shore islands, including Chinmen and Matsu regions. Training elementary and high school teachers as a nanotechnology human resources project and spreading the common knowledge of nanotechnology as the "seed teachers" to every corner of Taiwan has been accomplished.

In this program, there are fifteen principal target schools and fourteen assistant partnership schools working together. Two hundred and twenty-five teachers joined this program and are compiling the teaching material for the foundation of education. A total of eight people—one program chief host, two program collective hosts, and five program coordination hosts—control the direction of this program.

The executive results of the Nano-Technology K–12 Human Resources Development Program are divided into two main parts.

The first part—invisible: The training of the "seed teachers," learning and collaborating from different principal target schools; teachers attend special lectures of nanoscience and develop a common concept for curriculum from the lectures.

The second part—visible: Teachers develop comic books for students in elementary schools and animated cartoon for students in elementary and junior high school.

K–12 Nanotechnology Education Curriculum Project by Teachers

1. **Develop comic books for students in elementary schools**—A comic book entitled Nano BlasterMan was created for the elementary school students. It depicts the adventures of a superhero, Nano BlasterMan, who could use the nanotechnology to fight evil. The comic book was

suggested by teachers, drawn by a professional illustrator, and confirmed by the engineering faculty for technical details.

2. **Developing animated cartoons for students in elementary and junior high school**—The animated film entitled *A Fantastic Journey for Nana and Nono* was released in July 2004. It introduced the basic theory of nanotechnology and applications for daily life from a child's perspective. In addition, the NHRD Program Office also completed one interactive multimedia entitled *"Nano Magic"* in December 2005. Both of them have Chinese/English/Japanese versions now.
3. **Developing textbooks for students in senior high schools**—A three-volume set of books titled *Nanotechnology Symphony—Physics, Chemistry, and Biology,* was written by high school teachers as introductory material about nanotechnology. These books contained nanotechnology basic concepts, such as nano-dimension, nanomaterial, nano-catalyst, photonic crystal, and various applications of nanotechnology. It contained six experiments designed to give students hands-on experiences in a regular high school laboratory. The experiments included topics to cover synthesis of aqueous ferrofluid and diffraction of laser beams with ferrofluid.
4. **Developing a reference book for K–12 teachers and university students**—In order to train the seed teachers of the leading schools, a series of courses containing the basic and applied topics of nanoscience and technology were arranged from February to May 2004. There were fifteen scheduled courses during the period. The contents of the training courses were organized in the beginning to cover basic nanoscience and its application in different aspects. Experts specialized in each topics were invited to give a three-hour lesson. The contents of the courses were later collected to publish a book of *Nanotechnology—Fundamental, Application and Experiment.*
5. **Setting up AFM labs and mobile museum**—In order to provide the teachers and students with experiences of advanced nanotechnology equipment, the regional centers set up AFM labs at the participating universities, starting in September 2004. Since then, the AFM equipment was set up and operation manuals were developed gradually for K–12 teachers and students. To reach schools in remote areas, the main program office worked with the National Taiwan Science Education Center in Taipei to devise a traveling demonstration van in 2004. Domestic products, cloth, tile, and tennis racket nets using nanomaterials were displayed. Animated films, AFM scanner, and hands-on activities, light penetration and reflection of buckyballs were included. The van was staffed by volunteer teachers to provide on-site help at schools.

6. *Nano superman*—A full-color 3D comic book. This comic book was designed for the elementary school student to reinforce the understanding of nano knowledge.
7. **Hosting annual conference for seed teachers**—The conference is to demonstrate the results of the seed teachers after attending a series of training courses. The "seed teachers" present their own lesson plans and interchange ideas with others.
8. **Hosting nano summer camps for teachers and students**—During the camps, experts, professors, and experienced teachers gave talks and led hands-on activities. Topics included the science and technology background for nanotechnology and nano materials. Hands-on activities included topics such as "Making carbon nanotube models," "Making nano solar cells," etc. The purpose is to help the teachers and students learn about nanotechnology.
9. **Magic nano phenomenon in natural world**—Introduced the basic theory of nanotechnology and applications for daily life from a child's perspective.
10. **Game of Technology Midas–"Expert of Nano Technology"**—Using the Midas idea to teach the adults and children understanding of nanotechnology. The entertainment included the nanoscience lessons.
11. **E-Learning teaching material—"NM magic house"**—"NM magic house" was developed with the main program office working with the National Taiwan Science Education Center in Taipei. It contains "Nano Magic Books" (included ten animations), "The Laboratory of Super Doctor" (five movies and animation), "AFM," "Turntable" (included six games).

The Mid-North Region K–12 Education Center for Nanotechnology

1. **Nanotechnology education on-line resources**—The online resources, including "Nanotechnology Topic Maps," "Nanotechnology On-line Dictionary," and "NanoWiki" system Web site, integrate powerful tools for nanotechnology data base system.
2. **Hosting seminar and training courses for K–12 seed teachers**—The seminar, training courses, and labs tour aim to strengthen and broaden nanotechnology science for K–12 seed teachers.

3. **Extracurricular activities for nanotechnology education**—To promote nanotechnology science to students, the seed schools arrange many kinds of extracurricular activities, such as nanotechnology science camps, AFM/SEM labs tour, nanotechnology poster competitions, and so on, to attract students' attentions.
4. **Nanotechnology education publications**—To assist seed teachers on promoting nanotechnology science, to enhance students' understanding about nanotechnology, and to popularize nanotechnology science, a series of nanotechnology-related publications were scheduled to be published by senior seed teachers with the support from the Mid-North Region K–12 Education Center for Nanotechnology.

The Mid-South Region K–12 Education Center for Nanotechnology

1. **The Wonderland of Nanotechnology (interactive multimedia teaching disc)**—Titled "The Wonderland of Nanotechnology" this teaching disc comes from the style of on-line games that are very popular with the younger generation. The idea is to represent the introduction of nanotechnology and incorporate exploration of science activity. Designed to teach new technology through the students' favorite style of games and to inspire the desire to consider further study of nanotechnology.
2. **Science fiction battle—Chibi with Nanotechnology (animated teaching disc)**—After Zhuge Liang creates the Eastern wind, then he orders General Zhou Yu to use fire to attack Cao Cao's ship. At this moment, two meteor-like objects go through the sky, and Zhuge Liang realizes something is going to occur. One of the meteor-like objects goes to Cao's camp and the other to the Liao's camp. They are two researchers of nanotechnology who take a ride in time machines back to the period of the Three Kingdoms. Then a completely different "The Battle of Chibi" begins as follows ….
3. **Pork game for nano-photocatalyst**—The idea is to learn the function of nano-photocatalyst through the pork game. We design a very fashionable style of pork (not sure what pork means in their culture) and hope our young generation would learn new technology through the game.
4. **Training handbook for atomic force microscope**—This training handbook for an atomic force microscope (AFM) is specially designed for the teachers in K–12 programs. The contents of this handbook include the basic theory and the operation and application of AFM.

The South Region K–12 Education Center for Nanotechnology

Relevant Publications

1. ***Nanotechnology—Fundamental, Applications and Experiments***—This book is written by 16 university professors for training the seed teachers of leading schools. It is also suitable for the graduate and undergraduate students. The content is divided into three parts: (1) fundamentals (4 chapters), (2) applications (9 special topics), and (3) experiments (8 experiments). The publisher of this book is Gau Lih Book Co. (February 2005).
2. **Nanotechnology teaching material of Southern Center for K–12 Nanotechnology**—A three-volume book set entitled *Nanotechnology Teaching Material* was written by "seed teachers" and revised by renowned professors. As a professional instructional material, it introduces nano concepts and applications. With abundant pictures, it is quite appropriate for both students and teachers.
3. ***The Tiny but Beautiful Nano World***—The book is the first Braille material addressing nanoscience designed for blind students. The easy, comprehensible phrases and contents in the book will help blind students to also experience the magic of nanotechnology.
4. ***Nano Experiments Video***—Since the experimental setup for demonstrating experiments takes a lot of time, Southern Center decided to present the whole experiment including the preparation by video. There are eight experiments included in this video.
5. ***Nano Hands-On Teaching Video***—To expand nano education, Southern Center collaborated with Lu-Chu Senior High School to make a film presenting hands-on activities conducted in the Nano Science Camp for students.
6. ***Nanotechnology Teaching in Sign Language Video***—To popularize nanotechnology among disadvantaged minorities, the program instructor of 2006 headed for Kaohsiung County Special Education School (for disabled students) to give a lecture on nano education. During the lecture, hands-on experiments were also conducted. A professional sign language interpreter was invited to give simultaneous translations. Southern Center videotaped the whole lecture, and made it the first nanotechnology teaching material video in sign language.

The Major Activities

1. **Expanding nanotechnology education to remote areas**—Accessing the latest information, software, and hardware is difficult for schools

in remote areas. Therefore, expanding nanotechnology education to remote areas is one of the major activities of Southern Center.

2. **Holding the nano science camp for K–12 students**—In addition to instructing students on basic concepts of nanotechnology with multimedia materials, teachers also conducted hands-on activities to enhance students' nanotechnology knowledge in the camp.
3. **AFM lab training courses for seed teachers**—The training courses are divided into two parts: first, the lecture on theory and principles of mechanics and tools, and second, the operation of the mechanics and tools. There were 15 participants in 2005, 60 participants in 2006, and 86 participants in 2007. In 2008, Southern Center plans to continue with this activity, and professors will also be invited to instruct the application in AFM education and case study.

East Region Nanotechnology K–12 Education and Development Center

National Dong Hwa University has joined the national Nanotechnology K–12 Education and Development project since August 2003. The East Region Center works with the researchers and the K–12 teachers to promote learning and teaching nanotechnology in Hualien and Taitung county. Subject to the geography and demography of this region, we try to reach out, get the local teachers and students involved, think from their perspectives and bring in the resources they need, and adapt and compose the teaching materials and booklets.

The main activities are:

1. **Teacher education and cultivation**
   a. K–12 nanotechnology forums (2004–2007)
   b. Nanotechnology teaching materials and methods workshops (2006–2007)
   c. Nanotechnology atomic force microscopy (AFM) hands-on practice and experiment workshops (2005–2006)
   d. Symposiums of K–12 nanotechnology partner schools (2005–2007)
2. **Design teaching material and supporting multimedia**
   a. Nanotechnology multimedia teaching material workshops (2004–2007)
   b. Producing multimedia and VCD/DVD for teaching nanotechnology (2005–2006)

   c. Nanotechnology-oriented teachers' manual: teachers' manual for physics and chemistry teachers in high schools, teachers' manual for biology teachers in high schools.
   d. Easy accessible booklets: nanotechnology "Kung-Fu" booklets (2005–2007).
3. **Promotion of nanoeducation: reaching out and get involved**
   a. Project investigators visiting regional K–12 schools for project promotion (2005–2006)
   b. Join the educational science activities—A stand and posters at the 45th national high school science exhibition (held in Hualien) (2005)
   c. Junior and senior high school students visiting the nanoscience laboratory in National Dong Hwa University (2004–2007)
4. **Evaluation and reflection**
   a. Survey studies of the scientific literacy of the East Region K–12 nano teachers and students (2004–2007)
   b. Qualitative studies and analysis of East Region K–12 nano seed teacher teams (2004–2007)
   c. A documentary film of the East Region K–12 nanotechnology education promotion (2005–2007)

Become a "Seed Teacher" and Start the Process in the United States

If you decide to become a "seed teacher" in the United States, much of the curriculum is already available online with links provided in Part 3: Resources. The questions below may guide you toward exciting classroom experiences with your students.

Start by Asking Why? and Why Not?

- Can teachers learn the material on their own from the resources?
- If science is based on inquiry, why can it not be taught as inquiry?
- If you do not know the answer to a question a student might ask, why not turn it into inquiry, working together with one of the resource Web sites provided?
- Many of the materials developed for early grades are available under informal science resources. Why not use them to get started?

- Teachers are not expected to be scientists, so why not try the nanoscience games with your students to get started?
- Have fun with the resources and learn together. Rethink education together "from the bottom up."

References*

1. http://www.whitehouse.gov/photos-and-video/video/president-obama-hampton-university
2. http://www.reinventingschools.org/learning/
3. http://ies.ed.gov/ncee/wwc/
4. Dobbie, W., & Fryer, R. G., Jr., Are high-quality schools enough to close the achievement gap? Evidence from a social experiment in Harlem. (NBER Working Paper No. 15473). Cambridge, MA: National Bureau of Economic Research, 2009.
5. Judith Light Feather. The 1st International Collaboration in U.S. on K–12 nano science courses, 2009, http://www.tntg.org/documents/52.html
6. http://nano-1.colife.org.tw/en/entry/content!contentView.htm?id=11256

* All links active as of August 2010.

Index

V

W

Y

Z